Praise for
Why Darwin Matters

"A readable and well-researched book on what is perhaps the most vital scientific topic of our age. . . . Genial and intellectually uncompromising."
—Steven Pinker

"Thoughtfully explains why intelligent design is both bad science and poor religion, how a wealth of scientific data from varied fields support evolution, and why religion and science need not be in conflict. . . . Shermer's wit and passion will appeal to many."
—*Publishers Weekly*

"Expertly marshals point-by-point explanations of why evolution is worthwhile science, why ID isn't science at all, why ID criticisms of evolution are irrelevant, why science cannot invalidate religion, and why Christians and conservatives ought to accept evolution."
—*Booklist*

"Shermer is one of America's necessary minds. A reformed fundamentalist who is now an experienced foe of pseudoscience and superstition, he does us the double favor of explaining exactly what creationists believe, and then of demonstrating that they have no case. With his forensic and polemical skill, he could have left them for dead: instead he generously urges them to stop wasting their time (and ours) and do some real work."
—Christopher Hitchens

"The idea that evolution and God should be at odds is among the strangest of doctrines, an attempt to make the divine follow our particular notions of how He should operate. Shermer explains what really happened, in terms that should be accessible to any faithful reader."   —Bill McKibben

"Shermer offers calm, generally civil answers to the major questions about evolution, squarely faces controversy, generally forgoes cheap shots at the opposition. . . . A valuable, clearly presented tool in a key modern controversy."   —*Kirkus Reviews*

WHY DARWIN MATTERS

MICHAEL SHERMER

WHY DARWIN
MATTERS

THE CASE
AGAINST
INTELLIGENT
DESIGN

✦

A HOLT PAPERBACK

Henry Holt and Company   •   New York

Holt Paperbacks
Henry Holt and Company, LLC
Publishers since 1866
175 Fifth Avenue
New York, New York 10010
www.henryholt.com

A Holt Paperback® and ® are registered trademarks of
Henry Holt and Company, LLC.

Distributed in Canada by H. B. Fenn and Company Ltd.

For further information on the Skeptics Society and *Skeptic* magazine,
contact P.O. Box 338, Altadena, CA 91001, 626-794-3119;
fax: 626-794-1301; e-mail: skepticmag@aol.com.
www.skeptic.com

Library of Congress Cataloging-in-Publication Data
Shermer, Michael.
Why Darwin matters: the case against intelligent design /
Michael Shermer.—1st ed.
     p. cm.
ISBN-13: 978-0-8050-8306-4
ISBN-10: 0-8050-8306-5
     1. Evolution (Biology)   2. Intelligent design (Teleology)   I. Title.
QH366.2.S54 2006   2006041243
576.8—dc22

Henry Holt books are available for special promotions
and premiums. For details contact: Director, Special Markets.

Originally published in hardcover in 2006 by Times Books
First Holt Paperbacks Edition 2007
Designed by Victoria Hartman
Printed in the United States of America
5   7   9   10   8   6   4

To Frank J. Sulloway

"There ain't naught a man can't bear if he'll only be dogged.
It's dogged as does it."

In Darwin's footsteps in all ways

There is grandeur in this view of life, with its several powers,
having been originally breathed into a few forms or into one; and
that, whilst this planet has gone cycling on according to the fixed
law of gravity, from so simple a beginning endless forms most
beautiful and most wonderful have been, and are being, evolved.

—Charles Darwin, *On the Origin of Species,* 1859

CONTENTS

✦

PROLOGUE

✦

Why Evolution Matters

Hence both in space and time, we seem to be brought some-
what near to that great fact—that mystery of mysteries—the
first appearance of new beings on this earth.

—Charles Darwin, *Journal of Researches,* 1845

In June 2004, the science historian Frank Sulloway and I began a
month-long expedition to retrace Charles Darwin's footsteps in the
Galápagos Islands. It turned out to be one of the most physically
grueling experiences of my life, and as I have raced a bicycle across
America five times, that is saying something special about what the
young British naturalist was able to accomplish in 1835. Charles
Darwin was not only one sagacious scientist; he was also one tena-
cious explorer.[1]

I fully appreciated Darwin's doggedness when we hit the stark
and barren lava fields on the island of San Cristóbal, the first
place Darwin explored in the archipelago. With a sweltering equa-
torial sun and almost no fresh water, it is not long before water-
loaded seventy-pound packs begin to buckle your knees and strain
your back. Add hours of daily bushwhacking through dense,
scratchy vegetation, and the romance of fieldwork quickly fades. At
the end of one three-day excursion my water supply was so danger-
ously low that Frank and I collected the dew that had accumulated
on the tents the night before. One day I sliced my left shin on a
chunk of a'a lava. Another day I was stung by a wasp and one side

of my face nearly doubled in size. At the end of one particularly grueling climb through a moonscapelike area Darwin called the "craterized district," we collapsed in utter exhaustion, muscles quivering and sweat pouring off our hands and faces, after which we read from Darwin's diary, in which the naturalist described a similar excursion as "a long walk."

Death permeates these islands. Animal carcasses are scattered everywhere. The vegetation is coarse and scrappy. Dried and shriveled cactus trunks dot the bleak landscape. The lava terrain is so broken with razor-sharp edges that progress across it is glacially slow. Many people have died there, from stranded sailors of centuries past to wanderlust-driven tourists in recent years. Within days I had a deep sense of isolation and fragility. Without the protective blanket of civilization none of us are far from death. With precious little water and even less edible foliage, organisms eke out a precarious living, their adaptations to this harsh environment selected over millions of years. A lifelong observer of and participant in the evolution-creation controversy, I was struck by how clear it is in these islands: Creation by intelligent design is absurd. So how then did Darwin depart the Galápagos a creationist?

This is the question that Frank Sulloway went there to answer. Sulloway has spent a lifetime reconstructing how Darwin pieced together the theory of evolution. The iconic myth is that Darwin became an evolutionist in the Galápagos, discovering natural selection as he itemized finch beaks and tortoise carapaces, as he observed how each species had uniquely adapted to the available food and the island ecology. The legend endures, Sulloway notes, because it fits elegantly into a Joseph Campbell–like tripartite myth of the hero who (1) leaves home on a great adventure, (2) endures immeasurable hardship in the quest for noble truths, and (3) returns to deliver a deep message—in Darwin's case, evolution.

The myth is ubiquitous, appearing in everything from biology text-books to travel brochures, the latter of which inveigle potential customers to see what Darwin saw.

The Darwin Galápagos legend is emblematic of a broader myth that science proceeds by select eureka discoveries followed by sudden revolutionary revelations, as old theories fall before new facts. Not quite. Theories power perceptions. Nine months after departing the Galápagos, Darwin made the following entry in his ornithological catalogue about his mockingbird collection: "When I see these Islands in sight of each other, & possessed of but a scanty stock of animals, tenanted by these birds, but slightly differing in structure & filling the same place in Nature, I must suspect they are only varieties." He was seeing similar *varieties* of fixed kinds, not an *evolution* of separate species. Darwin did not even bother to record the island locations of the few finches he collected (and in some cases mislabeled), and these now-famous birds were never specifically mentioned in the *Origin of Species*. Darwin was still a creationist.[2]

Through careful analysis of Darwin's notes and journals, Sulloway dates Darwin's acceptance of evolution to the second week of March, 1837, after a meeting Darwin had with the eminent ornithologist John Gould, who had been studying Darwin's Galápagos bird specimens. With access to museum ornithological collections from areas of South America that Darwin had not visited, Gould corrected a number of taxonomic errors Darwin had made (such as labeling two finch species a "Wren" and an "Icterus"), and pointed out to him that although the land birds in the Galápagos were endemic to the islands, they were notably South American in character.

Darwin left the meeting with Gould, Sulloway concludes, convinced "beyond a doubt that transmutation must be responsible for the presence of similar but distinct species on the different islands of the Galápagos group." In Darwin's mind, the allegedly immutable "species barrier" had been shattered. That July, 1837, Darwin began his first notebook on *Transmutation of Species*. By 1844 he was confident enough to write in a letter to his botanist friend and colleague Joseph Hooker, "I was so struck with distribution of Galapagos organisms &c &c, & with the character of the American fossil mammifers &c &c, that I determined to collect blindly every sort of fact which cd bear any way on what are species." Five years at sea and nine years at home poring through "heaps" of books led Darwin to admit that, for him, "at last gleams of light have come, & I am almost convinced, (quite contrary to opinion I started with) that species are not (it is like confessing a murder) immutable."[3]

✦

Like confessing a murder. Dramatic words for something as seemingly innocuous as a technical problem in biology: the immutability of species. But it doesn't take a rocket scientist—or an English naturalist—to understand why the theory of the origin of species by means of natural selection would be so controversial: If new species are created naturally, what place, then, for God? No wonder Darwin waited twenty years before publishing his theory.[4]

From the time of Plato and Aristotle in ancient Greece to the time of Darwin and his fellow naturalist Alfred Russel Wallace in the nineteenth century, nearly everyone believed that a species retained a fixed and immutable "essence." A species, in fact, was defined by its very essence—the characteristics that made it like no other species. The theory of evolution by means of natural selection, then, is the theory of how kinds can become other kinds, and that upset not only the scientific cart, but the cultural horse pulling it. The great Harvard evolutionary biologist Ernst Mayr stressed just how radical was Darwin's theory: "The fixed, essentialistic species was the fortress to be stormed and destroyed; once this had been accomplished, evolutionary thinking rushed through the breach like a flood through a break in a dike."[5]

The dike, however, was slow to crumble. Darwin's close friend, the geologist Charles Lyell, withheld his support for a full nine years, and even then hinted at a providential design behind the whole scheme. The astronomer John Herschel called natural selection the "law of higgledy-piggledy." And Adam Sedgwick, a geologist and Anglican cleric, proclaimed that natural selection was a moral outrage, and penned this ripping harangue to Darwin:

> There is a moral or metaphysical part of nature as well as a physical. A man who denies this is deep in the mire of folly. You have ignored this link; and, if I do not mistake your meaning, you have done your best in one or two cases to break it. Were it possible (which thank God it is not) to break it, humanity, in my mind, would suffer a damage that might brutalize it, and sink the human race into a lower grade of degradation than any into which it has fallen since its written records tell us of its history.

In a review in *Macmillan's Magazine,* the political economist and social commentator Henry Fawcett wrote of the great divide surrounding the *Origin of Species:* "No scientific work that has been published within this century has excited so much general curiosity as the treatise of Mr. Darwin. It has for a time divided the scientific world with two great contending sections. A Darwinite and an anti-Darwinite are now the badges of opposed scientific parties."[6]

◆

Darwinites and anti-Darwinites. Although the scientific community is now united in agreement that evolution happened, a century and a half later the cultural world is still divided. According to a 2005 poll by the Pew Research Center, 42 percent of Americans hold strict creationist views that "living things have existed in their present form since the beginning of time" compared to 48 percent who believe that humans "evolved over time." Evolution has made news as the fight over teaching evolution has entered the courts and the school boards yet again. To that point, the Pew survey found that 64 percent said they were open to the idea of teaching creationism in addition to evolution in public schools, and more than half of those individuals said they think evolution should be *replaced* by creationism in biology classrooms.[7]

The evolution-creationism controversy is a cultural tempest in a scientific teapot—the debate is entirely cultural, even as professional scientists go about their business without giving Intelligent Design a second thought. Consider the geographic and political differences in attitudes about evolution, starting with the fact that evolution is under debate only in America (there are a few small creationist pockets in Australia, New Zealand, and the United Kingdom). And within the states, geography matters: 51 percent of Southerners accept the strict creationist view that humans were created as we are now and only 19 percent believe that we evolved through natural selection, while 59 percent of Northerners accept evolution through natural selection, and only 32 percent are creationists.

Given these demographics of belief, it came as no surprise to either conservatives or liberals when, in August 2005, President George W. Bush seemingly endorsed the teaching of Intelligent Design (ID) in public school science classes. As the story unfolded over the next two weeks, however, it became clear that the creationists, as well as many in the media and pundits on both the right and the left, had greatly exaggerated Bush's remarks. In an interview at the White House with a group of Texas newspaper reporters, Bush had said that when he was governor of Texas, "I felt like both sides ought to be properly taught." When a reporter asked for his position as president, Bush equivocated, saying, "I think that part of education is to expose people to different schools of thought. You're asking me whether or not people ought to be exposed to different ideas, and the answer is yes." Well, of course, but Bush answered a different question.

Indeed, Bush's science adviser, John H. Marburger III, said in a subsequent telephone interview with *The New York Times* that "evolution is the cornerstone of modern biology" and "intelligent design is not a scientific concept." He added that the president's comments should be interpreted to mean that ID might be discussed— not as science but as part of the "social context" in science classes, and that it would be "over-interpreting" Bush's remarks to conclude that the president believes that ID and evolution should be given equal treatment in public school science curricula.[8]

Rather than closing the controversy, Marburger's clarification helped stoke a renewed debate over whether evolution is "only a theory" and how it should be presented in the classroom. Indeed, in late 2005 the Kansas State Board of Education voted 6–4 to revise the state's science standards to include criticisms of evolution and to redefine science in a way that allows for the introduction of Intelligent Design creationism into the public school science curriculum (by deleting "natural explanations" from the definition of science). Shortly after the Kansas decision, a Bush-appointed conservative judge in Dover, Pennsylvania, ruled against Intelligent Design in a highly publicized court case. In early 2006 an Ohio board of education ruled not to include language that implies the introduction of Intelligent Design theory in science curricula. And there are at least a dozen more hot-spots around the country that will be settled by political debate, democratic vote, or a judge's decision.

But whatever happens in these politically charged skirmishes, truth in science is not determined by the *vox populi*. It does not matter whether 99 percent or just 1 percent of the public (or politicians) accepts a scientific theory—the theory stands or falls on the evidence, and there are few theories in science that are more robust than the theory of evolution. It took me a long time to realize this fact, for I began my career as a creationist. Saying this today almost feels like confessing a murder.

✦

Like confessing a murder. That is precisely how I felt when I realized that my creationist beliefs were wrong and that evolution actually happened. I became a creationist shortly after I became a born-again evangelical Christian in high school in 1971 and argued the creationist case through graduate school in 1977.[9] The evangelical movement was gathering momentum in the 1970s, and one of the central dogmas I took from it was that the biblical story of creation was to be taken literally; ergo, the theory of evolution had to be wrong.

Knowing next to nothing about evolution other than what I gleaned from reading creationist literature, I absorbed the arguments against the theory and practiced them on my undergraduate science and philosophy teachers. At Glendale College, which I attended for the first two years for general education requirements, I honed my debating skills as my creationist arguments were met with firm evolutionist counterarguments. At Pepperdine University, a Church of Christ institution where I finished my undergraduate degree, evolution was a nonentity as I witnessed for Christ and studied the theological underpinnings of the Christian faith. When I arrived at Pepperdine, in fact, I considered theology as a profession, but when I discovered that a doctorate required proficiency in Hebrew, Greek, Latin, and Aramaic, and knowing that foreign languages were not my strong suit (I struggled through two years of high school Spanish), I switched to psychology and mastered one of the languages of science: statistics. By the time I matriculated at

California State University at Fullerton for graduate training in experimental psychology, I was ensconced in the methods of science.

In science, the solutions to problems are based on established parameters to determine whether a hypothesis is probably right or definitely wrong. Statistics allow researchers to identify an event as likely to happen 99.99 percent of the time (rejecting the null hypothesis) or as insignificant. Instead of the rhetoric and disputation of theology, there are the logic and probabilities of science. What a difference this shift in thinking makes. In graduate school, I took a bevy of courses in research methods and statistics, and for recreation I signed up for a Tuesday evening course in evolution, just to see firsthand what had us creationists up in arms. The course was taught by an eccentrically charismatic biologist named Bayard Brattstrom, who from 7 to 10 P.M. regaled his class with breathtaking discoveries from the science of evolutionary biology, and who from 10 P.M. to closing time at the 301 Club just down the street held forth on science and religion, Darwin and Genesis, and all manner of related topics, accompanied by appropriate libations.

The scales fell from my eyes! It turned out that the creationist literature I was reading presented a Darwinian cardboard cutout that a child could knock down. (For example, if humans come from apes, why are apes still around? Of course, we didn't evolve from modern apes; apes and humans evolved from a common ancestor who lived nearly seven million years ago.) What I discovered was that the preponderance of evidence from numerous converging lines of scientific inquiry—geology, paleontology, zoology, botany, comparative anatomy, molecular biology, population genetics, biogeography, embryology, and others—all independently converge to the same conclusion: Evolution happened. *Why Darwin Matters* is about how we know evolution happened, in the context of the challenges to evolution mounted by twenty-first-century creationists and Intelligent Design theorists.

✦

Why does evolution matter? The influence of the theory of evolution on the general culture is so pervasive it can be summed up in a single observation: *We live in the age of Darwin.* Arguably the most culturally jarring theory in history, the theory of natural selection gave rise to the Darwinian revolution that changed both science and culture in ways immeasurable. On the scientific level, the static creationist model of species as fixed types was replaced with a fluid evolutionary model of species as ever-changing entities. The repercussions of this finding were, and are, astounding. The theory of top-down intelligent design of all life by or through a supernatural power was replaced with the theory of bottom-up natural design through natural forces. The anthropocentric view of humans as special creations placed by a divine hand above all others was replaced with the view of humans as just another animal species. The view of life and the cosmos as having direction and purpose from above was replaced with the view of the world as the product of the necessitating laws of nature and the contingent events of history. The view that human nature is infinitely malleable and primarily good was replaced with a view of human nature in which we are finitely restricted by our genes and are both good and evil.[10]

Darwin matters not only because his theory changed the world and reconfigured our position in nature, but because he launched a new and profound understanding of biology and science that has served future generations. Of the three intellectual giants of that epoch—Darwin, Marx, and Freud—only Darwin is still relevant for the simple reason that his theory was right, and the scientific evidence continues to support and refine it. In the memorable observation by geneticist Theodosius Dobzhansky, "Nothing in biology makes sense except in the light of evolution."[11]

1

THE FACTS OF EVOLUTION

$\leftrightarrow$

The affinities of all the beings of the same class have sometimes been represented by a great tree. I believe this simile
largely speaks the truth. As buds give rise by growth to fresh
buds, and these, if vigorous, branch out and overtop on all
sides many a feebler branch, so by generation I believe it has
been with the great Tree of Life, which fills with its dead and
broken branches the crust of the earth, and covers the surface
with its ever branching and beautiful ramifications.

—Charles Darwin, *On the Origin of Species,* 1859

The theory of evolution has been under attack since Charles Darwin first published *On the Origin of Species* in 1859. From the
start, its critics have seized on the *theory* of evolution to try to undermine its facts. But all great works of science are written in support of some particular view. In 1861, shortly after he published his
new theory, Darwin wrote a letter to his colleague, Henry Fawcett,
who had just attended a special meeting of the British Association
for the Advancement of Science during which Darwin's book was
debated. One of the naturalists had argued that *On the Origin of
Species* was too theoretical, that Darwin should have just "put his
facts before us and let them rest." In response, Darwin reflected
that science, to be of any service, required more than list-making;
it needed larger ideas that could make sense of piles of data.

Otherwise, Darwin said, a geologist "might as well go into a gravel-pit and count the pebbles and describe the colours."[1] Data without generalizations are useless; facts without explanatory principles are meaningless. A "theory" is not just someone's opinion or a wild guess made by some scientist. A theory is a well-supported and well-tested generalization that explains a set of observations. Science without theory is useless.

The process of science is fueled by what I call *Darwin's Dictum*, defined by Darwin himself in his letter to Fawcett: "all observation must be for or against some view if it is to be of any service."

Darwin's casual comment nearly a hundred and fifty years ago encapsulates a serious debate about the relative roles of data and theory, or observations and conclusions, in science.[2] In a science like evolution, in which inferences about the past must be made from scant data in the present, this debate has been exploded to encompass a fight between religion and science.

Prediction and Observation

Most essentially, *evolution is a historical science*. Darwin valued above all else prediction and verification by subsequent observation. In an act of brilliant historical science, for example, Darwin correctly developed a theory of coral reef evolution years before he developed his theory of biological evolution. He had never seen a coral reef, but during the *Beagle*'s famous voyage to the Galápagos, he had studied the types of coral reefs Charles Lyell described in *Principles of Geology*. Darwin reasoned that the different examples of coral reefs did not represent different types, each of which needed a different causal explanation; rather, the different examples represented different stages of development of coral reefs, for which only a single cause was needed. Darwin considered this a triumph of theory in driving scientific investigation: Theoretical prediction was followed by observational verification, whereby "I had therefore only to verify and extend my views by a careful examination of coral reefs."[3] In this case, the theory came first, then the data.

The publication of the *Origin of Species* triggered a roaring debate about the relative roles of data and theory in science. Darwin's "bulldog" defender, Thomas Henry Huxley, erupted in a paroxysm against those who pontificated on science but had never practiced it themselves: "There cannot be a doubt that the method of inquiry which Mr. Darwin has adopted is not only rigorously in accord with the canons of scientific logic, but that it is the only adequate method," Huxley wrote. Those "critics exclusively trained in classics or in mathematics, who have never determined a scientific fact in their lives by induction from experiment or observation, prate learnedly about Mr. Darwin's method," he bellowed, "which is not inductive enough, not Baconian enough, forsooth for them."[4]

Darwin insisted that theory comes to and from the facts, not from political or philosophical beliefs, whether from God or the godfather of scientific empiricism. It is a point he voiced succinctly in his cautions to a young scientist. The facts speak for themselves, he said, advising "the advantage, at present, of being very sparing in introducing theory in your papers; let theory guide your observations, but till your reputation is well established, be sparing of publishing theory. It makes persons doubt your observations."[5] Once Darwin's reputation was well established, he published his book that so well demonstrated the power of theory. As he noted in his autobiography, "some of my critics have said, 'Oh, he is a good observer, but has no power of reasoning.' I do not think that this can be true, for the *Origin of Species* is one long argument from the beginning to the end, and it has convinced not a few able men."[6]

Against Some View

Darwin's "one long argument" was with the theologian William Paley and the theory Paley posited in his 1802 book, *Natural Theology: or, Evidences of the Existence and Attributes of the Deity, Collected from the Appearances of Nature.* Sound eerily familiar? The scholarly agenda of this first brand of Intelligent Design was to correlate the works of God (nature) with the words of God (the Bible). Natural theology kicked off with John Ray's 1691 *Wisdom of God Manifested in Works of the Creation,* which itself was inspired by Psalms 19:11: "The Heavens declare the Glory of the Lord and the Firmament sheweth his handy work." John Ray, in what still stands as a playbook for creationism, explains the analogy between human and divine creations: If a "curious Edifice or machine" leads us to "infer the being and operation of some intelligent Architect or Engineer," shouldn't the same be said of "the Works of nature, that Grandeur and magnificence, that excellent contrivance for Beauty, Order, use, &c. which is observable in them, wherein they do as much transcend the Efforts of human Art and infinite Power and Wisdom exceeds finite" to make us "infer the existence and efficiency of an Omnipotent and All-wise Creator?"[7]

Paley advanced Ray's work through the accumulated knowledge of a century of scientific exploration. The opening passage of Paley's *Natural Theology* has become annealed into our culture as the winningly accessible and thus appealing "watchmaker" argument:

In crossing a heath, suppose I pitched my foot against a stone, and were asked how the stone came to be there. I might possibly answer, that, for any thing I knew to the contrary, it had lain there forever. But suppose I had found a watch upon the ground, and it should be enquired how the watch happened to be in that place. The inference, we think, is inevitable; that the watch must have had a maker; that there must have existed, at some time and in some place or other, an artificer or artificers who formed it for the purpose which we find it actually to answer; who comprehended its construction, and designed its use.

But life is far more complex than a watch—so the design inference is even stronger!

There cannot be design without a designer; contrivance without a contriver. . . . The marks of design are too strong to be got over. Design must have had a designer. That designer must have been a person. That person is GOD.[8]

For longer than we have had the theory of evolution, we have had theologians arguing for Intelligent Design.

From Natural Theology to Natural Selection

After abandoning medical studies at Edinburgh University, Charles Darwin entered the University of Cambridge to study theology with the goal of becoming a Church of England cleric. Natural theology provided him with a socially acceptable excuse to study natural history, his true passion. It also educated Darwin in the arguments on design popularized by Paley and others.[9] His intimacy with their ideas was respectful, not combative. For example, in November 1859, the same month that the *Origin of Species* was published, Darwin wrote his friend John Lubbock, "I do not think I hardly ever admired a book more than Paley's 'Natural Theology.' I could almost formerly have said it by heart."[10] Both Paley and Darwin addressed a problem in nature: the origin of the design of life. Paley's answer was to posit a top-down designer—God. Darwin's answer was to posit a bottom-up designer—natural selection. Natural theologians took this to mean that evolution was an attack on God, without giving much thought to what evolution is.

Ever since Darwin, much has been written about what, exactly, evolution is. Ernst Mayr, arguably the greatest evolutionary theorist since Darwin, offers a subtly technical definition: "evolution is change in the adaptation and in the diversity of populations of organisms." He notes that evolution has a dual nature, a "'vertical' phenomenon of adaptive change," which describes how a species responds to its environment over time, and a "'horizontal' phenomenon of populations, incipient species, and new species," which describes adaptations that break through the genetic divide.[11] And I'll never forget Mayr's definition of a species, because I had to memorize it in my first course on evolutionary biology: "A species is a group of actually or potentially interbreeding natural populations reproductively isolated from other such populations."[12]

Mayr outlines five general tenets of evolutionary theory that have been discovered in the years since Darwin published his revolutionary book:

1. *Evolution*: Organisms change through time. Both the fossil record of life's history and nature today document and reveal this change.
2. *Descent with modification*: Evolution proceeds through the branching of common descent. As every parent and child knows, offspring are similar to but not exact replicas of their

parents, producing the necessary variation that allows adaptation to the ever-changing environment.

3. *Gradualism:* All this change is slow, steady, and stately. Given enough time, small changes within a species can accumulate into large changes that create new species; that is, macroevolution is the cumulative effect of microevolution.

4. *Multiplication:* Evolution does not just produce new species; it produces an increasing number of new species.

And, of course,

5. *Natural selection:* Evolutionary change is not haphazard and random; it follows a selective process. Codiscovered by Darwin and the naturalist Alfred Russel Wallace, natural selection operates under five rules:

A. Populations tend to increase indefinitely in a geometric ratio: 2, 4, 8, 16, 32, 64, 128, 256, 512, 1024 . . .

B. In a natural environment, however, population numbers must stabilize at a certain level. The population cannot increase to infinity—the earth is just not big enough.

C. Therefore, there must be a "struggle for existence." Not all of the organisms produced can survive.

D. There is variation in every species.

E. Therefore, in the struggle for existence, those individuals with variations that are better adapted to the environment leave behind more offspring than individuals that are less well adapted. This is known as *differential reproductive success.*

As Darwin said, "as more individuals are produced than can possibly survive, there must in every case be a struggle for existence, either one individual with another of the same species, or with the individuals of distinct species, or with the physical conditions of life."[13]

The process of natural selection, when carried out over countless generations, gradually leads varieties of species to develop into new species. Darwin explained:

> It may be said that natural selection is daily and hourly scrutinising, throughout the world, every variation, even the slightest; rejecting that which is bad, preserving and adding up all that is good; silently and insensibly working, whenever and wherever opportunity offers, at the improvement of each organic being in relation to its organic and inorganic conditions of life. We see nothing of these slow changes in progress, until the hand of time has marked the long lapses of ages, and then so imperfect is our view into long past geological ages, that we only see that the forms of life are now different from what they formerly were.[14]

The time frame is long and the changes from generation to generation are subtle. This may be one of the most important and difficult points to grasp about the theory of evolution. It is tempting to see species as they exist today as a living monument to evolution, to condense evolution into the incorrect but provocative shorthand that humans descended from chimpanzees—a shorthand that undercuts the facts of evolution.

Natural selection is the process of organisms struggling to survive and reproduce, with the result of propagating their genes into the next generation. As such, it operates primarily at the local level. The Oxford evolutionary biologist Richard Dawkins elegantly described the process as "random mutation plus non-random cumulative selection,"[15] emphasizing the non-random. Evolution is not the equivalent of a warehouse full of parts randomly assorting themselves into a jumbo jet, as the creationists like to argue. If evolution were truly random there would be no biological jumbo jets.

Genetic mutations and the mixing of parental genes in offspring may be random, but the selection of genes through the survival of their hosts is anything but random. Out of this process of self-organized directional selection emerge complexity and diversity.

Natural selection is a description of a process, not a force. No one is "selecting" organisms for survival or extinction, in the benign sense of dog breeders selecting for desirable traits in show breeds, or in the malignant sense of Nazis selecting prisoners at Auschwitz-Birkenau. Natural selection, and thus evolution, is unconscious and nonprescient—it cannot look forward to anticipate what changes are going to be needed for survival. The evolutionary watchmaker is blind, says Dawkins, *pace* Paley.

By way of example, once when my young daughter asked how evolution works, I used the polar bear as an example of a "transitional species" between land mammals and marine mammals, because although they are land mammals they spend so much time in the water that they have acquired many adaptations to an aquatic life. But this is not correct. It implies that polar bears are *on their way* (in transition) to becoming marine mammals. They aren't. Polar bears are not "becoming" anything. Polar bears are well adapted for their lifestyle. That's all. If global warming continues, perhaps polar bears will adapt to a full-time aquatic existence, or perhaps they will move south and become smaller brown bears, or perhaps they will go extinct. Who knows? No one.

Where Are All the Fossils?

Evolution is a historical science, and historical data—fossils—are often the evidence most cited for and against it. In the creationist textbook, *Of Pandas and People*—one of the bones of contention in the 2005 Intelligent Design trial of *Kitzmiller et al. v. Dover Area School District,* in Dover, Pennsylvania—the authors state: "Design theories suggest that various forms of life began with their distinctive features already intact: fish with fins and scales, birds with feathers and wings, mammals with fur and mammary glands. . . . Might not gaps exist . . . not because large numbers of transitional forms mysteriously failed to fossilize, but because they never existed?"[16]

Darwin himself commented on this lack of transitional fossils, asking, "Why then is not every geological formation and every stratum full of such intermediate links?" In contemplating the answer, he turned to the data and noted that "geology assuredly does not reveal any such finely graduated organic chain; and this, perhaps, is the gravest objection which can be urged against my theory."[17] So where *are* all the fossils?

One answer to Darwin's dilemma is the exceptionally low probability of any dead animal's escaping the jaws and stomachs of predators, scavengers, and detritus feeders, reaching the stage of fossilization, and then somehow finding its way back to the surface through geological forces and unpredictable events to be discovered millions of years later by the handful of paleontologists looking for its traces. Given this reality, it is remarkable that we have as many fossils as we do.

There is another explanation for the missing fossils. Ernst Mayr outlines the most common way that a species gives rise to a new species: when a small group (the "founder" population) breaks away and becomes geographically—and thus reproductively—isolated from its ancestral group. As long as it remains small and detached, the founder group can experience fairly rapid genetic changes, especially relative to large populations, which tend to sustain their genetic homogeneity through diverse interbreeding.

Mayr's theory, called *allopatric speciation,* helps to explain why so few fossils would exist for these animals.

The evolutionary theorists Niles Eldredge and Stephen Jay Gould took Mayr's observations about how new species emerge and applied them to the fossil record, finding that gaps in the fossil record are not missing evidence of gradual changes; they are extant evidence of punctuated changes. They called this theory *punctuated equilibrium.*[18] Species are so static and enduring that they leave plenty of fossils in the strata while they are in their stable state (equilibrium). The change from one species to another, however, happens relatively quickly on a geological time scale, and in these smaller, geographically isolated population groups (punctuated). In fact, species change happens so rapidly that few "transitional" carcasses create fossils to record the change. Eldredge and Gould conclude that "breaks in the fossil record are real; they express the way in which evolution occurs, not the fragments of an imperfect record."[19] Of course, the small group will also be reproducing, following the geometric increases that are observed in all species, and will eventually form a relatively large population of individuals that retain their phenotype for a considerable time—and leave behind many well-preserved fossils. Millions of years later this process results in a fossil record that records mostly the equilibrium. The punctuation is in the blanks.

The Evidence of Evolution

In August 1996, NASA announced that it discovered life on Mars. The evidence was the Allan Hills 84001 rock, believed to have been ejected out of Mars by a meteor impact millions of years ago, which then fell into an orbit that brought it to Earth.[20] On the panel of NASA experts was paleobiologist William Schopf, a specialist in ancient microbial life. Schopf was skeptical of NASA's claim because, he said, the four "lines of evidence" claimed to support the find did not converge toward a single conclusion. Instead, they pointed to several possible conclusions.[21]

Schopf's analysis of "lines of evidence" reflects a method of science first described by the nineteenth-century philosopher of science William Whewell. To prove a theory, Whewell believed, one must have more than one induction, more than a single generalization drawn from specific facts. One must have multiple inductions that converge upon one another, independently but in conjunction. Whewell said that if these inductions "jump together" it strengthens the plausibility of a theory: "Accordingly the cases in which inductions from classes of facts altogether different have thus jumped together, belong only to the best established theories which the history of science contains. And, as I shall have occasion to refer to this particular feature in their evidence, I will take the liberty of describing it by a particular phrase; and will term it the Consilience of Inductions."[22] I call it a *convergence of evidence*.

Just as detectives employ the convergence of evidence technique to deduce who most likely committed a crime, scientists employ the method to deduce the likeliest explanation for a particular phenomenon. Cosmologists reconstruct the history of the universe through a convergence of evidence from astronomy, planetary geology, and physics. Geologists reconstruct the history of the planet through a convergence of evidence from geology, physics, and chemistry. Archaeologists piece together the history of civilization through a convergence of evidence from biology (pollen grains), chemistry (kitchen middens), physics (potsherds, tools), history (works of art, written sources), and other site-specific artifacts.

As a historical science, evolution is confirmed by the fact that so many independent lines of evidence converge to its single conclusion. Independent sets of data from geology, paleontology, botany, zoology, herpetology, entomology, biogeography, comparative anatomy and physiology, genetics and population genetics, and many other sciences each point to the conclusion that life evolved. This is a convergence of evidence. Creationists can demand "just one fossil transitional form" that shows evolution. But evolution is not proved through a single fossil. It is proved through a convergence of fossils, along with a convergence of genetic comparisons between species, and a convergence of anatomical and physiological comparisons between species, and many other lines of inquiry. For creationists to disprove evolution, they need to unravel all these independent lines of evidence, as well as construct a rival theory that can explain them better than the theory of evolution. They have yet to do so.

The Tests of Evolution

Creationists like to argue that evolution is not a science because no one was there to observe it and there are no experiments to run today to test it. The inability to observe past events or set up controlled experiments is no obstacle to a sound science of cosmology, geology, or archaeology, so why should it be for a sound science of evolution? The key is the ability to test one's hypothesis. There are a number of ways to do so, starting with the broadest method of how we know evolution happened.

Consider the evolution of our best friend, the dog. With so many breeds of dogs popular for so many thousands of years, one would think that there would be an abundance of transitional fossils providing paleontologists with copious data from which to reconstruct their evolutionary ancestry. Not so. In fact, according to Jennifer A. Leonard of the National Museum of Natural History in Washington, D.C., "the fossil record from wolves to dogs is pretty sparse."[23] Then how do we know the origin of dogs? In a 2002 issue of *Science,* Leonard and her colleagues report that mitochrondrial DNA (mtDNA) data from early dog remains "strongly support the hypothesis that ancient American and Eurasian domestic dogs share a common origin from Old World gray wolves." In the same issue of *Science,* Peter Savolainen from the Royal Institute of Technology in Stockholm and his colleagues note that the fossil record is problematic "because of the difficulty in discriminating between small wolves and domestic dogs," but their study of mtDNA sequence variation among 654 domestic dogs from around the world "points to an origin of the domestic dog in East Asia ~15,000 yr B.P." from a single gene pool of wolves. Finally, Brian Hare from Harvard and his colleagues describe the results of their study in which they found that domestic dogs are more skillful than wolves at using human communicative signals indicating the location of hidden food, but that "dogs and wolves do not perform differently in a non-social memory task, ruling out the possibility that dogs outperform wolves in all human-guided tasks." Therefore, "dogs' social-communicative skills with humans were acquired during the process of domestication."[24] Although no single fossil proves that dogs came from wolves, the convergence of evidence from archaeological, morphological, genetic, and behavioral "fossils" reveals the ancestor of all dogs to be the East Asian wolf.

The tale of human evolution is revealed in a similar manner (although here we do have an abundance of transitional fossil riches), as it is for all ancestors in the history of life. One of the finest compilations of evolutionary convergence is Richard Dawkins's magnum opus, *The Ancestor's Tale,* 673 pages of convergent science recounted with literary elegance. Dawkins traces innumerable "transitional fossils" (what he calls "concestors"—the "point of rendezvous" of the last common ancestor shared by a set of species) from *Homo sapiens* back four billion years to the origin of replicating molecules and the emergence of evolution. No one concestor proves that evolution happened, but together they reveal a majestic story of a process over time.[25] We know human evolution happened because innumerable bits of data from myriad fields of science conjoin to paint a rich portrait of life's pilgrimage.

But the convergence of evidence is just the start. The *comparative method* allows us to infer evolutionary relationships using data from a wide variety of fields. Luigi Luca Cavalli-Sforza and his colleagues, for example, compared fifty years of data from population genetics, geography, ecology, archaeology, physical anthropology, and linguistics to trace the evolution of the human races. Using both the convergence and comparative methods led them to conclude that "the major stereotypes, all based on skin color, hair color and form, and facial traits, reflect superficial differences that are not confirmed by deeper analysis with more reliable genetic traits." By comparing surface (physical) traits—the phenotype of individuals—with genetic traits—the genotype—they teased out the relationship between different groups of people. Most interesting, they found that the genetic traits disclosed "recent evolution mostly under the effect of climate and perhaps sexual selection." For example, they discovered that Australian aborigines are genetically more closely related to southeast Asians than they are to African blacks, which makes sense from the perspective of the evolutionary timeline: The migration pattern of humans out of Africa would have led them first to Asia and then to Australia.[26]

Dating techniques provide evidence of the timeline of evolution.

The dating of fossils, along with the earth, moon, sun, solar system, and universe, are all tests of evolutionary theory, and so far they have passed all the tests. We know that the earth is approximately 4.6 billion years old because of the convergence of evidence from several methods of dating rocks: Uranium Lead, Rubidium Strontium, and Carbon-14. Further, the age of the earth, the age of the moon, the age of the sun, the age of the solar system, and the age of the universe are consistent, maintaining yet another consilience. If, say, the earth was dated at 4.6 billion years old but the solar system was dated at one million years old, the theory of evolution would be in trouble. But Uranium Lead, Rubidium Strontium, and Carbon-14 have not provided any good news for the so-called Young Earth creationists.

Better yet, the fossils and organisms speak for themselves. *Fossils do show intermediate stages,* despite their rarity. For example, there are now at least eight intermediate fossil stages identified in the evolution of whales. In human evolution, there are at least a dozen known intermediate fossil stages since hominids branched off from the great apes six million years ago. *And geological strata consistently reveal the same sequence of fossils.* A quick and simple way to debunk the theory of evolution would be to find a fossil horse in the same geological stratum as a trilobite. According to evolutionary theory, trilobites and mammals are separated by hundreds of millions of years. If such a fossil juxtaposition occurred, and it was not the product of some geological anomaly (such as uplifted, broken, bent, or even flipped strata—all of which occur but are traceable), it would mean that there was something seriously wrong with the theory of evolution.

Evolution also posits that *modern organisms should show a variety of structures from simple to complex, reflecting an evolutionary history rather than an instantaneous creation.* The human eye, for example, is the result of a long and complex pathway that goes back hundreds of millions of years. Initially a simple eyespot with a handful of light-sensitive cells that provided information to the organism about an important source of the light, it developed into a recessed eyespot, where a small surface indentation filled with light-sensitive cells provided additional data on the direction of light; then into a deep recession eyespot, where additional cells at greater depth provide more accurate information about the environment; then into a pinhole camera eye that is able to focus an image on the back of a deeply recessed layer of light-sensitive cells; then into a pinhole lens eye that is able to focus the image; then into a complex eye found in such modern mammals as humans. All of these structures are expressed in modern eyes.

Further, *biological structures show signs of natural design.* The anatomy of the human eye, in fact, shows anything but "intelligence" in its design. It is built upside down and backwards, requiring photons of light to travel through the cornea, lens, aqueous fluid, blood vessels, ganglion cells, amacrine cells, horizontal cells, and bipolar cells before they reach the light-sensitive rods and cones that transduce the light signal into neural impulses—which are then sent to the visual cortex at the back of the brain for processing into meaningful patterns. For optimal vision, why would an intelligent designer have built an eye upside down and backwards? This "design" makes sense only if natural selection built eyes from available materials, and in the particular configuration of the ancestral organism's pre-existing organic structures. The eye shows the pathways of evolutionary history, not of intelligent design.

Additionally, *vestigial structures stand as evidence of the mistakes, the misstarts, and, especially, the leftover traces of evolutionary history.* The cretaceous snake *Pachyrhachis problematicus,* for example, had small hind limbs used for locomotion that it inherited from its quadrupedal ancestors, gone in today's snakes. Modern whales retain a tiny pelvis for hind legs that existed in their land mammal ancestors but have disappeared today. Likewise, there are wings on flightless birds, and of course humans are replete with useless vestigial structures, a distinctive sign of our evolutionary ancestry. A short list of just ten vestigial structures in humans leaves one musing: Why would an Intelligent Designer have created these?

1. *Male nipples.* Men have nipples because females need them, and the overall architecture of the human body is more efficiently developed in the uterus from a single developmental structure.

2. *Male uterus.* Men have the remnant of an undeveloped female reproductive organ that hangs off the prostate gland for the same reason.

3. *Thirteenth rib.* Most modern humans have twelve sets of ribs, but 8 percent of us have a thirteenth set, just like chimpanzees and gorillas. This is a remnant of our primate ancestry: We share common ancestors with chimps and gorillas, and the thirteenth set of ribs has been retained from when our lineage branched off six million years ago.

4. *Coccyx.* The human tailbone is all that remains from our common ancestors' tails, which were used for grasping branches and maintaining balance.

5. *Wisdom teeth.* Before stone tools, weapons, and fire, hominids were primarily vegetarians, and as such we chewed a lot of plants, requiring an extra set of grinding molars. Many people still have them, despite the smaller size of our modern jaws.

6. *Appendix.* This muscular tube connected to the large intestine was once used for digesting cellulose in our largely vegetarian diet before we became meat eaters.

7. *Body hair.* We are sometimes called "the naked ape"; however, most humans have a layer of fine body hair, again left over from our evolutionary ancestry from thick-haired apes and hominids.

8. *Goose bumps.* Our body hair ancestry can also be inferred from the fact that we retain the ability of our ancestors to puff up their fur for heat insulation, or as a threat gesture to potential predators. Erector pili—"goose bumps"—are a telltale sign of our evolutionary ancestry.

9. *Extrinsic ear muscles.* If you can wiggle your ears you can thank our primate ancestors, who evolved the ability to move their ears independently of their heads as a more efficient means of discriminating precise sound directionality and location.

10. *Third eyelid.* Many animals have a nictitating membrane that covers the eye for added protection; we retain this "third eyelid" in the corner of our eye as a tiny fold of flesh.

Evolutionary scientists can provide dozens more examples of vestigial structures—let alone examples of how we know evolution happened from all of these other various lines of historical evidence. Yet as a science, evolution depends primarily on the ability to test a hypothesis. How can we ever test an evolutionary hypothesis if we cannot go into a lab and create a new species naturally?

I once had the opportunity to help dig up a dinosaur with Jack Horner, the curator of paleontology at the Museum of the Rockies in Bozeman, Montana. As Horner explains in his book *Digging Dinosaurs,* "paleontology is not an experimental science; it's an historical science. This means that paleontologists are seldom able to test their hypotheses by laboratory experiments, but they can still test them."[27] Horner discusses this process of historical science at the famous dig in which he exposed the first dinosaur eggs ever found in North America. The initial stage of the dig was "getting the fossils out of the ground." Unsheathing the bones from the overlying and surrounding stone is backbreaking work. As you move from jackhammers and pickaxes to dental tools and small brushes, historical interpretation accelerates as a function of the rate of bone unearthed. Then, in the second phase of a dig, he gets "to look at the fossils, study them, make hypotheses based on what we saw and try to prove or disprove them."

When I arrived at Horner's camp I expected to find the busy director of a fully sponsored dig barking out orders to his staff. I was surprised to come upon a patient historical scientist, sitting cross-legged before a cervical vertebra from a 140-million-year-old *Apatosaurus* (formerly known as *Brontosaurus*), wondering what to make of it. Soon a reporter from a local paper arrived inquiring of Horner what this discovery meant for the history of dinosaurs. Did it change any of his theories? Where was the head? Was there more than one body at this site? Horner's answers were those of a cautious scientist: "I don't know yet." "Beats me." "We need more evidence." "We'll have to wait and see." It was historical science at its best.

After two long days of exposing nothing but solid rock and my own ineptness at seeing bone within stone, one of the paleontologists pointed out that the rock I was about to toss away was a piece of bone that appeared to be part of a rib. If it was a rib, then the bone should retain its riblike shape as more of the overburden was chipped away. This it did for about a foot, until it suddenly flared to the right. Was it a rib, or something else? Horner moved in to check. "It could be part of the pelvis," he suggested. If it was part of the pelvis, then it should also flare out to the left when more was uncovered. Sure enough, Horner's prediction was verified by further digging.

In science, this process is called the *hypothetico-deductive method,* in which one forms a hypothesis based on existing data, deduces a prediction from the hypothesis, then tests the prediction against further data. For example, in 1981 Horner discovered a site in Montana that contained approximately thirty million fossil fragments of approximately ten thousand *Maiasaur*s in a bed measuring 1.25 miles by .25 miles. His hypothesizing began with a question: "What could such a deposit represent?"[28] There was no evidence that predators had chewed the bones, yet many were broken in half lengthwise. Further, the bones were all arranged from east to west—the long dimension of the bone deposit. Small bones had been separated from bigger bones, and there were no bones of baby *Maiasaur*s, only those of individuals between nine and twenty-three feet long. What would cause the bones to splinter lengthwise? Why would the small bones be separated from the big bones? Was this one giant herd, all killed at the same time, or was it a dying ground over many years?

An early hypothesis—that a mud flow buried the herd alive—was rejected because "it didn't make sense that even the most powerful flow of mud could break bones lengthwise . . . nor did it make sense that a herd of living animals buried in mud would end up with all their skeletons disarticulated." Horner constructed another hypothesis. "It seemed that there had to be a twofold event," he reasoned, "the dinosaurs dying in one incident and the bones being swept away in another." Since there was a layer of volcanic ash 1.5 feet above the bone bed, volcanic activity was implicated in the death of the herd. Horner then deduced that only fossil bones would split lengthwise, and therefore the damage to the bones had occurred long after the dying event. His hypothesis and deduction led to his conclusion that the herd was "killed by the gases, smoke and ash of a volcanic eruption. And if a huge eruption killed them all at once, then it might have also killed everything else around." Then perhaps there was a flood, maybe from a breached lake, carrying the rotting bodies downstream, separating the big bones from the small, lighter bones, and giving the bones a uniform orientation.[29]

A paleontological dig is a good example of how hypothetico-deductive reasoning and historical sciences can make predictions based on initial data that are then verified or rejected by later historical evidence. Evolutionary theory is rooted in a rich array of data from the past that, while nonreplicable in a laboratory, are nevertheless valid sources of information that can be used to piece together specific events and test general hypotheses. While the specifics of evolution—how quickly it happens, what triggers species change, at which level of the organism it occurs—are still being studied and unraveled, the general theory of evolution is the most tested in science over the past century and a half. Scientists agree: Evolution happened.

2

WHY PEOPLE DO NOT
ACCEPT EVOLUTION

———————— ✦ ————————

The real attack of evolution, it will be seen, is not upon ortho-
dox Christianity or even upon Christianity, but upon religion—the most basic fact in man's existence and the most
practical thing in life. If taken seriously and made the basis of
a philosophy of life, it would eliminate love and carry man
back to a struggle of tooth and claw.

—William Jennings Bryan, closing statement, Scopes trial, 1925

On a thickly muggy and stiflingly hot summer day in 1925, William
Jennings Bryan arose to speak at the end of the trial of the century.
He had carefully crafted his speech to express the deeper question
he felt had been put on trial in Dayton, Tennessee: not, as most observers believed, whether a high school science teacher had lec-
tured his students on Darwin's theory of evolution, but rather who
would win the battle for humanity's soul. "The soul is immortal and
religion deals with the soul," he wrote in a statement epigrammati-
cally poignant; "the logical effect of the evolutionary hypothesis is
to undermine religion and thus affect the soul."

Bryan was never actually allowed to deliver his dramatic final
speech in the Scopes "Monkey Trial." The judge determined that it
was irrelevant to the case—the same ruling he made against the defense when they called evolutionary biologists as expert witnesses—

and Bryan died rather unceremoniously two days after the trial's end. But the speech was subsequently published as a booklet heroically entitled *Bryan's Last Speech: The Most Powerful Argument against Evolution Ever Made.*[1] The speech is an insight into why so many people resist the theory of evolution: the belief, and fear, that accepting evolution leads to the breakdown of morality and the loss of meaning for humanity. The syllogistic reasoning goes as follows:

> *Evolution implies that there is no God, therefore . . .*

> *Belief in the theory of evolution leads to atheism, therefore . . .*

> *Without a belief in God there can be no morality or meaning, therefore . . .*

> *Without morality and meaning there is no basis for a civil society, therefore . . .*

> *Without a civil society we will be reduced to living like brute animals.*

This is what bothers people about evolutionary theory, not the technical details of the science. Most folks don't give one whit about adaptive radiation, allopatric speciation, phenotypic variation, assortative mating, allometry and heterochrony, adaptation and exaptation, gradualism and punctuated equilibrium, and the like. What they do care about is whether teaching evolution will make their kids reject God, allow criminals and sinners to blame their genes for their actions, and generally cause society to fall apart.

Where did they get such an idea?

The Real Legacy of the Scopes Trial

Forget the Lindbergh kidnapping trial, the Manson murder trial, even the O.J. media trial. The Scopes trial really did beat them all as a test of our humanity. It was bigger than life, from the issues at hand to the characters involved.[2]

It helps to know that the trial was initially instigated as a publicity stunt, dreamed up by the fledgling American Civil Liberties Union in collaboration with the city leaders of the economically struggling Tennessee town of Dayton. On one side of the dock was the most famous defense attorney of his era, Clarence Darrow; on the other was the century's preeminent orator and defender of the faith, three-time presidential candidate Bryan. Covering the trial for the *Baltimore Sun* was the unapologetically cynical reporter H. L. Mencken, who meted out such barbs as this: "If the Anti-Evolutionists in Tennessee were aware of the existence of any other religions than their own, they might realize that it is the very genius of religion itself to evolve from primary forms to higher forms. The author of the anti-evolution bill is obviously nearer in mental development to the nomads of early biblical times than he is to the intelligence of the young man who is under trial."[3]

That young man, John Thomas Scopes, was a substitute teacher from a neighboring county who, by his own admission, volunteered to challenge Tennessee's "anti-evolution" law because, in addition to being a freethinker, he thought that the extended stay in Dayton over the summer and the ensuing attention might help his cause in a local love interest. The ACLU was sure they would lose in Dayton—giving them a chance to appeal to the Tennessee State Supreme Court and eventually land the case in the U.S. Supreme

Court. From the start, court cases about teaching evolution have been about everything *but* the science.

Most people think that science scored a knockout victory in Tennessee. Reading Mencken would certainly lead to this conclusion. Of Bryan he gibed: "Once he had one leg in the White House and the nation trembled under his roars. Now he is a tinpot pope in the Coca-Cola belt and a brother to the forlorn pastors who belabor half-wits in galvanized iron tabernacles behind the railroad yards. . . . It is a tragedy, indeed, to begin life as a hero and to end it as a buffoon." In fact, this was no victory for evolution or science, and it may surprise readers to learn that Scopes's guilty verdict was overturned not on the merits of the case but on a minor technicality involving the levying of a fine of over $50 by a judge instead of a jury. Embarrassed by the bad publicity the state of Tennessee was receiving, the state legislators used a technical misstep to prevent the case from reaching the state's supreme court. Who can blame them after reading comments like this from Mencken: "It serves notice on the country that Neanderthal man is organizing in these forlorn backwaters of the land, led by a fanatic, rid of sense and devoid of conscience." Worse, the controversy stirred by the trial made textbook publishers and state boards of education reluctant to deal with evolution in any manner. A study of high school biology textbooks before and after the trial revealed that the subject of evolution simply disappeared from the curriculum and was not taught for decades.[4]

Bryan's story reveals a common fear many people hold about the theory of evolution. A liberal and freethinker on so many other issues, Bryan took a stand against evolutionary theory after the First World War, when he became aware of the use of social Darwinism justifying militarism, imperialism, eugenics, and what he saw as "paralyzing the hope of reform" through its program of "scientific breeding, a system under which a few supposedly superior intellects, self-appointed, would direct the mating and the movements of the mass of mankind." He developed this view after reading the entomologist Vernon L. Kellog's 1917 book *Headquarters Nights,* a recounting of the evenings Kellog spent listening to German military and intellectual leaders justify their militarism and imperialistic expansionism with classic social Darwinism—national survival of the fittest, improvement of the superior Germanic breed, and elimination of unfit races.[5]

Bryan became concerned for both his faith and his country. The enemy he identified, however, was not Germany, but evolutionary theory. "The evolutionary hypothesis carried to its logical conclusion, disputes every vital truth of the Bible," he wrote in his final speech. "Its tendency, naturally, if not inevitably, is to lead those who really accept it, first to agnosticism and then to atheism. Evolutionists attack the truth of the Bible, not openly at first, but by using weasel-words like 'poetical,' 'symbolical,' and 'allegorical' to search out the meaning of the inspired record of man's creation." Scopes's crime was to pass this poison on to the next generation:

> The people of Tennessee have been patient enough; they acted none too soon. How can they expect to protect society, and even the church, from the deadening influence of agnosticism and atheism if they permit the teachers employed by taxation to poison the mind of the youth with this destructive doctrine? And remember, that the law has not heretofore required the writing of the word "poison" on poisonous doctrines. The bodies of our people are so valuable that the druggists and physicians must be careful to properly label all poisons; why not be as careful to protect the spiritual life of our people from the poisons that kill the soul?

Bryan's fears about social Darwinism were rankled by the lawyer across the aisle. He narrowed his focus on Darrow, particularly on the attorney's famous and very public defense of Nathan Leopold and Richard Loeb, the teenagers who abducted fourteen-year-old Bobby Franks and clubbed him to death with a chisel in what they had thought would be the "perfect crime." After the boys confessed to the murder, Darrow agreed to take the case, employing a defense to shift the penalty from death to life in prison. His was a deterministic view of human behavior. "Man is in no sense the maker of himself and has no more power than any other machine to escape the law of cause and effect," Darrow opined. The boys were not ultimately responsible for the murder because human volition is a fiction: "each act, criminal or otherwise, follows a cause; that given the same conditions the same result will follow forever and ever."[6] Darrow claimed that Leopold and Loeb were themselves victims, and their trial served as a platform for Darrow to argue the larger case that our actions are the product of environmental influences.[7] Now Darrow was defending evolution in Dayton, and Bryan foresaw the future in which lawyers could argue that we are all just products of our brute animal heritage, "coerced by a fate fixed by the laws of heredity," and thus not morally culpable for our actions. Bryan would not stand for it: If evolution were accepted, it would "destroy all sense of responsibility and menace the morals of the world."

This was the "Great Commoner" who engaged great causes, the man who famously defended labor and attacked the gold standard by declaring, "you shall not press down upon the brow of labor this crown of thorns, you shall not crucify mankind upon a cross of gold."[8] Of course Bryan saw the teaching of evolution as a war between science and religion:

> Evolution is at war with religion because religion is supernatural, it is therefore the relentless foe of Christianity which is a revealed religion. Let us, then, hear the conclusion of the whole matter. Science is a magnificent material for force, but is not a

teacher of morals. It can also build gigantic intellectual ships, but it constructs no moral rudders for control of storm tossed human vessels. It not only fails to supply the spiritual element needed, but some of its unproven hypotheses rob the ship of its compass and thus endanger its cargo.

Soaring prose from a towering man, but he claimed a false war between science and religion. Accepting evolution does not force us to jettison our morals and ethics, and rejecting evolution does not ensure their constancy. *We should not press down upon the brow of education this crown of religious thorns; we should not crucify science upon a cross of religious gold.*

The Search for Truth

"Darwin's bulldog," Thomas Henry Huxley, proclaimed that the *Origin of Species* was "the most potent instrument for the extension of the realm of knowledge which has come into man's hands since Newton's *Principia*," and lamented to himself "how extremely stupid not to have thought of that!" Ernst Mayr asserted that "it would be difficult to refute the claim that the Darwinian revolution was the greatest of all intellectual revolutions in the history of mankind." Stephen Jay Gould called the theory of evolution one of the half dozen most important ideas in the entire history of Western thought. Richard Dawkins inquired what common ground we could find for conversation with an extraterrestrial intelligence, and answered "evolution"—because it is "a universal truth" that is common throughout the cosmos.[9]

If the theory of evolution is so profound and proven, why doesn't everyone accept it as true? One source of resistance is the confusion over the verbs "accept" and "believe." I use the verb "accept" instead of the more common expression "believe in" because evolution is not a religious tenet, to which one swears allegiance or belief as a matter of faith. It is a factual reality of the empirical world. Just as one would not say "I believe in gravity," one should not proclaim "I believe in evolution." But getting hung up on the idea that one is supposed to "believe in" evolution just as you "believe in" God is just one brand of resistance to evolution. There are at least five specific reasons people resist the truth of evolutionary theory:

1. *A general resistance to science.* If you imagine that you have to "believe in" a scientific theory, a conflict arises between science and religion, in which you are forced to choose one over the other. In particular, if scientific discoveries do not appear to support religious tenets, the religious tend to opt for religion, while the secular tend to opt for science.

2. *Belief that evolution is a threat to specific religious tenets.* Occasionally, rather than choosing religion over science, religious believers attempt to use science to prove religious tenets, or to mold scientific findings to fit religious beliefs. For example, the effort to prove that the Genesis creation story is accurately reflected in the geological fossil record has led many creationists to conclude that the earth was created within the past 10,000 years. This is in sharp contrast to the geological evidence for a 4.6-billion-year-old Earth. If one insists on the findings of science squaring true with religious doctrines, this can lead to conflict between science and religion.

3. *The fear that evolution degrades our humanity.* After Copernicus toppled the pedestal of our cosmic centrality, Darwin delivered the coup de grâce by revealing us to be "mere" animals, subject to the same natural laws and historical forces as all other animals.

Copernicus no longer generates controversy because his theory of heliocentrism is about the relative place and position of cosmic real estate, whereas Darwin's theory remains controversial because it is about us, which we take personally.

4. *The equation of evolution with ethical nihilism and moral degeneration.* Decrying the inevitable dark hole of existence that comes from a "life without meaning" has become a potent tool of social persuasion—what "meaning" means, precisely, being left to whoever is leading the lament. The neoconservative social commentator Irving Kristol expressed the sentiment tidily in 1991: "If there is one indisputable fact about the human condition it is that no community can survive if it is persuaded—or even if it suspects—that its members are leading meaningless lives in a meaningless universe."[10] Similar fears were raised by Nancy Pearcey, a fellow of the Intelligent Design hothouse the Discovery Institute, in a briefing before the House Judiciary Committee of the U.S. Congress. Pearcey cited a popular song urging "you and me, baby, ain't nothing but mammals so let's do it like they do on the Discovery Channel" and went on to claim that since the American legal system is based on moral principles, the only way to generate ultimate moral grounding—and, apparently, change the Billboard Top 40—is for the law to have an "unjudged judge," an "uncreated creator."[11] God and mammon in the halls of Congress.

5. *The fear that evolutionary theory implies we have a fixed human nature.* The first four reasons for the resistance to evolutionary theory come almost exclusively from the political right. This last reason originates from the political left, from liberals who fear that the application of evolutionary theory to human thought and action implies that political policy and economic doctrines will fail because the constitution of humanity is stronger than the

constitutions of states. (This is what I call "liberal creationism," the doppelganger of conservative creationism.)[12]

These fears say nothing about the evidence for evolution, and they do no more than conjecture about the effects of accepting evolution on human psychology and behavior. There will always be some—True Believers—who, like Bryan, will not be able to set aside their biggest fear: that accepting a scientific view of the natural world will challenge their faith in God.

But if one is a theist, it should not matter *when* God made the universe—ten thousand years ago or ten billion years ago. The difference of six zeros is meaningless to an omniscient and omnipotent being, and the glory of divine creation cries out for praise regardless of when it happened. Likewise, it should not matter *how* God created life—whether it was through a miraculous spoken word or through the natural forces of the universe that He created. The grandeur of God's works command awe regardless of what processes He used. We have learned a lot in four thousand years, and that knowledge should never be dreaded or denied. Theists and theologians should embrace science, especially evolutionary theory, for what it has done to reveal the magnificence of the divinity in a depth never dreamed by our ancient ancestors.

The Greater Threat

There is, however, a greater threat to the theory of evolution today: not from those who resist evolution, but from those who misunderstand it. Most people know very little about evolution, and this makes it easier for the people who do not accept evolution to encourage others to question the theory, even to the point of denial.[13] In a 2001 Gallup poll, for example, a quarter of the people surveyed said they didn't know enough to say whether they accepted evolution or not, and only 34 percent considered themselves to be "very informed" about it. Because evolution is so controversial, public school science teachers typically drop the subject entirely rather than face the discomfort aroused among students and parents. What is not taught is not learned.[14]

The modern Intelligent Design movement has seized on this misunderstanding, from their claim that evolution is "only a theory" to their narrowing of the scientific method to experiments in a laboratory to their insistence that any appearance of order in the natural world proves both design and the existence of a supernatural designer. It is this last argument that is especially appealing to those who are unsure about what exactly evolution means: What is so wrong with teaching that there is an intelligent design to life? Is there something wrong with wanting to see an Intelligent Designer in the universe?

3

IN SEARCH OF THE DESIGNER

———————————— ◆ ————————————

> Man has been here 32,000 years. That it took a hundred mil-
> lion years to prepare the world for him is proof that that is what
> it was done for. I suppose it is. I dunno. If the Eiffel tower were
> now representing the world's age, the skin of paint on the pin-
> nacle-knob at its summit would represent man's share of that
> age; and anybody would perceive that that skin was what the
> tower was built for. I reckon they would, I dunno.
>
> —Mark Twain, "Was the World Made for Man?" 1903

Why do you believe in God?

I have been asking people this question for most of my adult
life. In 1998, Frank Sulloway and I presented the query in a more
official format—along with the question "Why do you think *other
people* believe in God?"—in a survey given to ten thousand Americans. Just a few of the answers we received:

> A 22-year-old male law student with moderate religious convic-
> tions (a self-rated five on a nine-point scale), who was raised by
> very religious parents and who today calls himself a deist, writes,
> "I believe in a creator because there seems to be no other pos-
> sible explanation for the existence of the universe," yet other
> "people believe in God to give their lives purpose and meaning."

> A 43-year-old female physician with a self-rated three on the
> nine-point scale of religious conviction says she believes in God

because "I experience peace and serenity in myself and my life. It is always there when I choose to allow the experience of it. This is proof to me of divine intelligence, a Higher Power, the Oneness of all," yet "I think most people believe in God because they are taught to. It is, in fact, a *belief* rather than a direct experience for them."

A 43-year-old male computer scientist and Catholic with very strong religious convictions (a nine on the nine-point scale) "had a personal conversion experience, where I had direct contact with God. This conversion experience, and ongoing contacts in prayer, form the only basis for my faith." Other people believe in God, however, "because of (a) their upbringing, (b) the comfort of the church, and (c) a hope for this contact."

A 36-year-old male journalist and evangelical Christian with a self-rated eight in religious conviction writes: "I believe in God because to me there is ample evidence for the existence of an intelligent designer of the universe." Yet, "others accept God out of a purely emotional need for comfort throughout their life and use little of their intellectual capacity to examine the faith to which they adhere."

A 40-year-old female Catholic nurse with very strong religious convictions (a nine on the nine-point scale) says that "I believe in God because of the example of my spiritual teacher who believes in God and has unconditional love for people and gives so completely of himself for the good of others. And since I have followed this path, I now treat others so much better." On the other hand, she writes that "I think people initially believe in God because of their parents and unless they start on their own path—where they put a lot of effort into their spiritual part of their life—they continue to believe out of fear."

Searching for Answers

When Sulloway and I noticed the difference between why people believe in God and why they think *other* people believe in God, we decided to undertake an extensive analysis of all the written answers people provided in our survey. In addition, we inquired about family demographics, religious background, personality characteristics, and other factors that contribute to religious belief and skepticism. We discovered that the seven strongest predictors of belief in God are:

1. being raised in a religious manner
2. parents' religiosity
3. lower levels of education
4. being female
5. a large family
6. lack of conflict with parents
7. being younger

In sum, being female and raised by religious parents in a large family appears to make one more religious, whereas being male, educated, in conflict with one's parents, and older appears to make one less religious.[1] As people become older and more educated, they encounter other belief systems that lead them to see the connection between various personal and social influences and religious beliefs. This helps explain the differences we observed in reasons people give for their own beliefs versus the reasons they attribute to other people's beliefs.

From the responses we received in a preliminary survey, we created a taxonomy of eleven categories of reasons people give for their own and others' beliefs. The five most common answers given to the question *Why do you believe in God?*:

1. The good design / natural beauty / perfection / complexity of the world or universe (28.6%)
2. The experience of God in everyday life (20.6%)
3. Belief in God is comforting, relieving, consoling, and gives meaning and purpose to life (10.3%)
4. The Bible says so (9.8%)
5. Just because / faith / the need to believe in something (8.2%)

And the six most common answers given to the question *Why do you think other people believe in God?*:

1. Belief in God is comforting, relieving, consoling, and gives meaning and purpose to life (26.3%)
2. Religious people have been raised to believe in God (22.4%)
3. The experience of God in everyday life (16.2%)
4. Just because / faith / the need to believe in something (13.0%)
5. Fear death and the unknown (9.1%)
6. The good design / natural beauty / perfection / complexity of the world or universe (6.0%)

Notice that the intellectually based reasons offered for belief in God—"the good design of the universe" and "the experience of God in everyday life"—which occupied first and second place when people were describing their own beliefs dropped to sixth and third place, respectively, when they were describing the beliefs of others. Indeed, when reflecting on others' beliefs, the two most common reasons cited were emotion-based (and fear-averse!):

personal comfort ("comforting, relieving, consoling") and social comfort ("raised to believe").

Sulloway and I believe that these results are evidence of an *intellectual attribution bias,* in which people consider their own beliefs as being rationally motivated, whereas they see the beliefs of others as being emotionally driven. By analogy, one's commitment to a political belief is generally attributed to a rational decision ("I am for gun control because statistics show that crime decreases when gun ownership decreases"), whereas another person's opinion on the same subject is attributed to need or emotional reasons ("he is for gun control because he is a bleeding-heart liberal who needs to identify with the victim"). This intellectual attribution bias appears to be equal opportunity on the subject of God. The apparent good design of the universe, and the perceived action of a higher intelligence in daily activities, are powerful intellectual justifications for belief. But we readily attribute other people's belief in God to their emotional needs and how they were raised.

Designed to Find Design?

The intellectual attribution bias may be the result of evolution. Perceiving the world as well designed and thus the product of a designer, and even seeing divine providence in the daily affairs of life, may be the product of a brain adapted to finding patterns in nature. We are pattern-seeking as well as pattern-finding animals. One of numerous studies that supports this supposition was an experiment conducted by Stuart Vyse and Ruth Heltzer in which subjects participated in a video game. The goal of the game was to navigate the path of a cursor through a matrix grid using directional keys. One group of subjects were awarded points when they successfully found a way through the grid's lower right portion, while a second group of subjects were awarded points randomly. Both groups were subsequently asked to describe how they thought the points were awarded. Most of the subjects in the first group found the pattern of point scoring and accurately described it. Similarly, most of the subjects in the second group also found "patterns" of point scoring, even though no such patterns existed.[2]

Finding patterns in nature may have an evolutionary explanation: There is a survival payoff for finding order instead of chaos in the world, and being able to separate threats (to fight or flee) from comforts (to embrace or eat, among other things), which enabled our ancestors to survive and reproduce. We are the descendants of the most successful pattern-seeking members of our species. In other words, we were designed by evolution to perceive design. How recursive!

Of course, until 1859 when Charles Darwin explained the natural, bottom-up origins of design, the default explanation—reverted to by most peoples in most cultures throughout most of history—was God. Since the most common reason people give for why they believe in God is the good design of the world, Intelligent Design creationists are tapping into the intuitive understanding most people hold about life and the universe.

But there is a deep-seated flaw in this argument that undermines the entire endeavor. If the world is complex and looks intricately designed, and therefore the best inference is that there must be an intelligent designer, should we not then infer that an intelligent designer must itself have been designed? That is, if the earmarks of design imply that there is an intelligent designer, then the existence of an intelligent designer denotes that it must have a designer—a *super intelligent designer*. And by the same course of reasoning, any designer who can create a super intelligent designer must itself be a *superior super intelligent designer*.

Ad infinitum. Which brings us right back to the natural world, and the search for natural explanations for natural phenomena.

Shermer's Last Law: ID, ET, and God

One day I was thinking about what we might find if we went in search of an intelligent designer when I remembered Arthur C. Clarke's famous Third Law: "Any sufficiently advanced technology is indistinguishable from magic."[3] This led me to consider what a sufficiently advanced Extra-Terrestrial Intelligence (ETI) would be indistinguishable from, which led me to formulate *Shermer's Last Law: Any sufficiently advanced Extra-Terrestrial Intelligence is indistinguishable from God.*[4]

God is described by most Western religions as omniscient and omnipotent. Since we are far from the mark on these traits, how could we possibly distinguish a God who has them absolutely, from an Extra-Terrestrial Intelligence who has them in copious amounts relative to us? Thus, we would be unable to distinguish between absolute and relative omniscience and omnipotence. But if God were only relatively more knowing and powerful than us, then by definition He *would* be an ETI! From this I conclude that there is no difference between Intelligent Design, Extra-Terrestrial Intelligence, and God, at least a God that is part of our world. This conclusion is derived from the following sequence of observations and deductions:

Observation. Biological evolution is glacially slow compared to technological evolution. The reason is that biological evolution is Darwinian and requires generations of differential reproductive success, whereas technological evolution is Lamarckian, where change is inherited within the same generation.

Observation. The cosmos is very big and space is very empty, so the probability of making contact with an ETI is remote. By example, the speed of our most distant spacecraft, *Voyager I,* relative to the sun is 17.246 kilometers per second. The speed of light is 300,000 kilometers per second, so *Voyager I* is traveling at .0000574 percent of the speed of light. The Alpha Centauri star system, the closest to our sun, is 4.3 light-years away. This means that even traveling at a breakneck speed of 38,578 miles per hour, it would take *Voyager I* 74,912 years to get there (and it isn't even heading in that direction).[5]

Deduction. Ergo, the probability of making contact with an ETI who is only slightly more advanced than us is virtually nil. If we ever do encounter an ETI it will be as if a million-year-old *Homo erectus* were dropped into the middle of Manhattan, given a computer and a cell phone, and instructed to communicate with *Homo sapiens sapiens.* An ETI would be to us as we would be to this early hominid—godlike.

Observation. Science and technology have changed our world more in the past century than it changed in the previous hundred centuries—it took ten thousand years to get from the cart to the airplane, but only sixty-six years to get from powered flight to a lunar landing. Moore's Law of computer power doubling every eighteen months continues unabated and is now down to about a year. Some computer scientists, such as Ray Kurzweil, calculate that there have been thirty-two doublings since World War II, and that as early as 2030 we may encounter the Singularity—the point at which total computational power will rise to levels that are so far beyond anything that we can imagine that they will appear near infinite and thus, relatively speaking, be indistinguishable from omniscience (note the suffix!).[6] When this happens the world will change more in a decade than it did in the previous thousand decades.

Deduction. Extrapolate these trend lines out a hundred thousand years, or a million years (an eye blink on an evolutionary time scale, and thus a realistic estimate of how far advanced an ETI will be[7]), and we get a gut-wrenching, mind-warping feel for just how godlike an ETI would appear to us.

By pursuing a course of inquiry to its natural end, Intelligent Design, as a simplistic scientific explanation of the world around us, can only lead to the discovery of Extra-Terrestrial Intelligence. What Intelligent Design advocates will find (if they find anything) is an alien being capable of engineering DNA, cells, complex organisms, planets, stars, galaxies, and perhaps even universes. If today we can engineer genes, clone mammals, and manipulate stem cells with science and technologies developed in only the last half century, think of what an ETI could do with one hundred thousand years of equivalent powers of progress in science and technology. For an ETI who is a million years more advanced than we are, engineering the creation of planets and stars will probably be practicable. And if universes are created out of collapsing black holes—which some cosmologists think is probable—it is not inconceivable that a sufficiently advanced ETI could even create a universe.

What would we call an intelligent being that could engineer a universe, stars, planets, and life? If we knew the underlying science and technology used to do the engineering, we would call it an ETI; if we did not know the underlying science and technology, we would call it an Intelligent Designer; if we left science out of theology altogether, we would call it God.

So Intelligent Design is simple science. It is also bad theology: Intelligent Design reduces the deity to a mere engineer, a garage tinkerer, a technician piecing together worlds and life forms out of available materials, but not necessarily the creator of the original materials. This is not at all the spirit invoked in creeds such as those

formulated at the first Nicene Council, held in AD 325, the opening
of which states: "We believe in one God, the Father Almighty,
Maker of heaven and earth, and of all things visible and invisible."

In his classic work, *Maker of Heaven and Earth*, the Protestant
theologian Langdon Gilkey, who also penned a cogent history of
the 1981 Arkansas creationism trial,[8] rejects the approach taken
by the natural theologians all the way back to William Paley's hey-
day. "In the Christian doctrine of creation," he writes, "God is the
source of all and creates out of nothing. Thus the Christian idea,
far from merely representing a primitive anthropomorphic projec-
tion of human art upon the cosmos, systematically repudiates all
direct analogy from human art." Far from being a mere intelligent
watchmaker, God is the "transcendent source of all existence" who
creates *ex nihilo*—from nothing. For Gilkey, whose theology I
greatly respect, knowledge of God comes "not from a careful scientific or metaphysical analysis of the general experience of nature
and of finite existence, but rather from the illumination that comes
from special encounters with God in revelatory experiences."[9]

Recall that in our study on why people believe in God, the second most popular reason people gave for their belief was the experience of God in everyday life. This reason, and not the convoluted
logic and twisted science of Intelligent Design Creationism, makes
for deep and honest theology, the type of theology practiced by the
great German theologian Paul Tillich, who once said, "God does
not exist. He is being itself beyond essence and existence. Therefore to argue that God exists is to deny him."[10]

If we think of God as a thing, a being that exists in space and
time, it constrains God to our world, a world of other things and
other beings that are also restrained by the laws of nature and the
contingencies of chance. But if God is the maker of all things and
all beings visible and invisible in heaven and earth, God must be

above such restraints; that is, above the laws of nature and contingencies of chance. "The question of the existence of God can be neither asked nor answered," Tillich explains. "If asked, it is a question about that which by its very nature is above existence, and therefore the answer—whether negative or affirmative—implicitly denies the nature of God. It is as atheistic to affirm the existence of God as it is to deny it. God is being-itself, not *a* being."[11]

If there is a God, the avenue to Him is not through science and reason, but through faith and revelation. If there is a God, He will be so wholly Other that no science can reach Him, especially not the science that calls itself Intelligent Design.

4

DEBATING INTELLIGENT DESIGN

————————————◆————————————

Those who cavalierly reject the Theory of Evolution, as not adequately supported by facts, seem quite to forget that their own theory is supported by no facts at all.

—Herbert Spencer, *Essays Scientific,*
Political and Speculative, 1891

Debating creationists on the "science," rather than the theology, of Intelligent Design is problematic, and I have been encouraged not to do so by such friends and colleagues as Stephen Jay Gould and Richard Dawkins, whose wisdom I consider inestimable. They argue, correctly, that there is no debate about the reality of evolution—the issue was settled a century ago—and that the numerous debates about *how* evolution happened are all well within the normal borders of science. In any case, public debate is not how the validity of a scientific theory is determined, and participating in a debate could send the wrong message to the public: that evolution is debatable.

But eschewing public debate has not kept the Intelligent Design movement at bay, nor has the public decided that evolution should not be debated. On a balmy spring Southern California evening in 2004, for example, I entered the four-hundred-seat Physical Sciences Lecture Hall on the campus of the University of

California, Irvine, to find it chockablock with five hundred people. I was there to debate the incorrigibly insouciant Kent Hovind, Young Earth Creationist, Defender of the Fundamental Faith, and the fastest talker I have ever met. On the docket that evening was the defining question of this controversy, *Creation vs. Evolution: Creation (supernatural action) or Evolution (natural processes)—which is the better explanation?* So, in contrast to many other scientists, I believe there are legitimate and important reasons to engage in the debate between evolution and Intelligent Design:

1. Debates will occur anyway, so they might as well include someone with expertise and experience in science. Better still if they also have expertise and experience in debating, and can employ diplomacy, wit, and warmth along with scientific facts.

2. When a controversy receives as much media attention as evolution and creationism have garnered over the past century, refusing to engage in public dialogue or debate can be misconstrued as a weakness in one's position. Intelligent Design advocates have made a cottage industry out of exactly this point.

3. Debate forces both sides to put their cards face up on the table for everyone to see. Intelligent Design creationists have no science to speak of, and debates provide opportunities to demonstrate that a pretty PowerPoint slide is not science. As John Stuart Mill argued in his classic 1859 treatise *On Liberty:* "But the peculiar evil of silencing the expression of an opinion is, that it is robbing the human race; posterity as well as the existing generation; those who dissent from the opinion, still more than those who hold it. If the opinion is right, they are deprived of the opportunity of exchanging error for truth: if wrong, they lose, what is almost as great a benefit, the clearer perception and livelier impression of truth, produced by its collision with error."[1]

4. Debate is a chance to educate people about science and evolution. As we have discussed, muddled misunderstandings about the science of evolution—and the theology of Intelligent Design—far outnumber concrete understanding. In general, the world may be divided into three types of people: True Believers, Fence Sitters, and skeptics. Religious True Believers will never change their minds no matter what evidence is presented to them, and science-embracing skeptics already accept evolution. The battleground is for the Fence Sitters—those who have heard something about a claim or controversy and wonder what the explanation for it might be. Lacking a good explanation, the mind defaults to whatever explanation is on the table, regardless of how improbable it may be.

Point four is particularly relevant to the debate between evolution and Intelligent Design. Before the theory of evolutionary biology was developed in the nineteenth century, the default explanation for the distribution of species around the globe was independent creation by God and the Noachian flood. (Among the handful of more religiously skeptical scientists, the mode of distribution was Lamarckian evolution and long-gone land bridges between continents and islands.) But after Charles Darwin and Alfred Russel Wallace demonstrated how natural selection changes varieties into different species when they migrate into different regions, the default supernatural explanation could be abandoned in favor of a fact-based natural one. Debate affords us an occasion to demonstrate to Fence Sitters that there is, in fact, a perfectly reasonable natural explanation for the apparently supernatural phenomenon of design.[2]

Investigating Intelligent Design

Skepticism—thoughtful inquiry—is a scientific way of thinking; science—a testable body of knowledge—is applied skepticism. At the core of any scientific investigation are a number of skeptical principles that help us assess the validity of a claim before we examine the specific arguments and evidences. Six principles of skepticism help us sort through the various arguments for Intelligent Design:

Hume's Maxim, or, what is more likely?

In 1758 the Scottish philosopher David Hume published his most influential work, *An Enquiry Concerning Human Understanding.* When confronted by the claim of an event so improbable and extraordinary that it is called a miracle, Hume asks us to inquire what is more likely: that a supernatural act occurred contrary to the laws of nature, or that people who describe such acts are mistaken in their assessment of the event's supernatural nature? *Hume's Maxim* is best defined in his own words: "The plain consequence is (and it is a general maxim worthy of our attention), *'That no testimony is sufficient to establish a miracle, unless the testimony be of such a kind, that its falsehood would be more miraculous than the fact which it endeavours to establish.'* "

Hume then weighs which is more likely: "When anyone tells me that he saw a dead man restored to life, I immediately consider with myself whether it be more probable, that this person should either deceive or be deceived, or that the fact, which he relates, should really have happened. I weigh the one miracle against the other; and according to the superiority, which I discover, I pronounce my decision, and always reject the greater miracle. If the falsehood of his testimony would be more miraculous than the event which he relates; then, and not till then, can he pretend to command my belief or opinion."[3]

The evolution–Intelligent Design debate boils down to a Humean question of what's more likely: that the diversity and complexity of life we see around us came about by laws of nature that we can observe, or supernaturally by an Intelligent Designer that we cannot observe? The nineteenth-century social Darwinist Herbert Spencer answered the question rhetorically: "Well, which is the most rational theory about these ten millions of species? Is it most likely that there have been ten millions of special creations? or is it most likely that, by continual modifications due to change of circumstances, ten millions of varieties have been produced, as varieties are being produced still?"[4] It's a good question.

The Known and the Unknown, or, before you say
something is out of this world, first make sure it is not
in this world.

Creationists and Intelligent Design theorists explain natural phenomena by turning immediately to supernatural forces operating outside this world. Following that thinking, there is no point in searching for medical cures in the natural world; we should throw out all medical knowledge and turn purely to prayer, without trying any treatments. But medicine, like evolution and other sciences, deals with natural explanations for natural phenomena, and even with only a rudimentary understanding of the natural world we should seek answers from the known, natural world first. Before assuming that the unknown is the only possible way to account for the complexity and diversity of life, it must first be demonstrated that known forces and processes cannot do the job.

Burden of Proof, or, extraordinary claims require extraordinary evidence.

Darwin's original claim of evolution by means of natural selection was an extraordinary claim in its time, so he was required to provide extraordinary evidence for it. He did, and evidence has continued accumulating ever since. Today, the burden of proof is on creationists and Intelligent Design advocates to provide extraordinary evidence for their extraordinary claim that a supernatural being of great power and intelligence performed a supernatural act in place of or contrary to natural law. They have yet to do so.

Either-Or Fallacy, or, disproving A does not prove B.

The either-or fallacy is the false assumption that there are only two positions, A and B, so if A is wrong then B must be right. The fallacy is that discrediting A does not demonstrate B. Both A and B could be wrong and a third alternative could be correct. Creationists employ the either-or fallacy when they claim that life was either divinely created or naturally evolved. By attempting to discredit evolution they hope to draw the conclusion that creationism is true. In science, however, it is not enough to just debunk the accepted theory. You must also replace it with a theory that explains both the "normal" data accounted for by the old theory as well as some of the "anomalous" data not accounted for by the old theory. At the end of my debate with Kent Hovind, for example, he was asked by an audience member: "What is the best evidence for the creation?" He answered: "The impossibility of the contrary" (that is, evolution). In that simple statement, Hovind confessed the scientific sin of creationism and Intelligent Design: Disproving evolution does not prove creationism.

The Fossil Fallacy, or, one datum does not a science make.

In debates with creationists they often demand "just one transitional fossil" that proves evolution, pointing to a gap in the fossil record. When I fill the gap—for example, with *Ambulocetus natans,* a transitional fossil between ancient land mammals and modern whales—they respond that there are now *two gaps* in the fossil record! This is a clever retort, but it reveals a deep error in epistemology that I call the *Fossil Fallacy*: the belief that a single "fossil"—one bit of data—constitutes proof of a multifarious process or historical sequence.[5]

Proof is derived not through a single piece of evidence, but through that convergence of evidence from numerous lines of inquiry, all of which point to an unmistakable conclusion. We know evolution happened not because of a single transitional fossil such as *Ambulocetus natans,* but because of the convergence of evidence from such diverse fields as geology, paleontology, biogeography, comparative anatomy and physiology, molecular biology, genetics, and many more. Paleontologist Donald Prothero, for example, employs the convergence technique in revealing that in addition to at least eight transitional fossils from land mammals to whales, DNA from living specimens reveals that modern whales descended from even-toed hoofed mammals called *artiodactyls.* Whales, it turns out, are most closely related to the hippopotamus.[6]

No single discovery from any one field constitutes proof of evolution. It is the mass of data together, converging to reveal that life evolved in a specific sequence by a particular process, that makes the theory of evolution a singular landmark in our understanding of the world.

Methodological Naturalism, or, no miracles allowed.

In one of Sidney Harris's most poignant cartoons, two scientists are at a blackboard filled with equations, in which the words "THEN A MIRACLE OCCURS" appear in the middle. The caption has one scientist saying to the other: "I THINK YOU NEED TO BE MORE EXPLICIT HERE IN STEP TWO." This is what we call the *God of the Gaps* argument—wherever an apparent gap exists in scientific knowledge, this is where God interjects a miracle. Intelligent Design advocates argue something like this:

- X looks designed
- I can't think of how X could have been designed naturally
- Therefore X must have been designed supernaturally

This is comparable to the "plane problem" of Isaac Newton's time: The planets all lie in a plane (the plane of the ecliptic) and revolve about the sun in the same direction. Newton found this arrangement to be so improbable that he invoked God as an explanation at the end of his magisterial work *Principia Mathematica:* "This most beautiful system of the sun, planets, and comets could only proceed from the counsel and dominion of an intelligent and powerful Being."[7] Why don't creationists use this argument any more? Because astronomers have filled that gap with a natural explanation.

The technical term for this process is *methodological naturalism,* and Intelligent Design theorists harp about it incessantly. Methodological naturalism holds that life is the result of natural processes in a system of material causes and effects that does not allow, or need, the introduction of supernatural forces. In his book *Darwin on Trial,* University of California, Berkeley, law professor Phillip Johnson, the founding father of the Intelligent Design movement, accused scientists of unfairly defining God out of the picture by limiting the search to only natural causes, and argued that scientists who postulate that there are supernatural forces or interventions at work in the natural world are being pushed out of the scientific arena on the basis of nothing more than a fundamental rule of the game. Johnson wants the rules of the game changed to allow *methodological supernaturalism.*

Okay, let's change the rules. Let's allow methodological supernaturalism into science. What would that look like? How would that work? What would we do with supernaturalism? For the sake of argument, let's assume that Intelligent Design theorists have suddenly become curious about how exactly the Intelligent Designer operates. As researchers who are now given entrée into the scientific stadium with an addendum to the rules that allows supernaturalism, they call a time out during the game to announce, "THEN A MIRACLE OCCURS." What do we do now? Do we halt all future experiments? Since science is what scientists do, what are we supposed to *do* with such supernatural explanations? My response to the God of the Gaps argument is: "I THINK YOU NEED TO BE MORE EXPLICIT HERE IN STEP TWO."

Even if Intelligent Design advocates are willing to continue searching, what will they do if they discover a new force of nature that accounts for design? How will they identify it? Will it be considered a natural force, or a supernatural force? When electromagnetism and the weak and strong nuclear forces were discovered in the nineteenth and twentieth centuries, scientists did not identify them as supernatural forces; they simply added them to the known forces of nature. If IDers eschew all attempts to provide a naturalistic explanation for life, they abandon science altogether.[8]

There is no such thing as the supernatural or the paranormal. There is only the natural, the normal, and mysteries we have yet to explain.

Intelligent Design's Best Arguments

As we have seen, creationists and Intelligent Design theorists have made dozens of arguments trying to disprove evolution, most hinging on the truly meaningless search for a single piece of data that will fill the gap of the week. But much of their recent success in classrooms and with school boards has been in using the language of science to argue that the data support Intelligent Design rather than evolution. Here I present their ten most cogent—and most commonly presented—arguments, followed by an evolutionary response grounded in the latest scientific theories on the origin and evolution of the universe and life.[9]

The Anthropic Principle:
The universe is fine tuned for life.

We begin with what I consider to be the best scientific argument that creationists and Intelligent Design theorists have in their arsenal: The universe is finely tuned and delicately balanced to support life. Change any number of physical parameters or initial conditions of the universe by even the tiniest amount, and life would not be possible. Fine tuning implies that there is a fine tuner, an Intelligent Designer, a God.

There is no shortage of observations from leading scientists on this condition of the cosmos. No less a personage than Stephen Hawking wrote:

> Why is the universe so close to the dividing line between collapsing again and expanding indefinitely? In order to be as close as we are now, the rate of expansion early on had to be chosen fantastically accurately. If the rate of expansion one second after the big bang had been less by one part in 10^{10}, the universe would have

collapsed after a few million years. If it had been greater by one part in 10^{10}, the universe would have been essentially empty after a few million years. In neither case would it have lasted long enough for life to develop. Thus one either has to appeal to the anthropic principle or find some physical explanation of why the universe is the way it is.[10]

What is this supernaturally appealing "anthropic principle"? In *The Anthropic Cosmological Principle,* the physicists John Barrow and Frank Tipler define the term: "It is not only man that is adapted to the universe. The universe is adapted to man. Imagine a universe in which one or another of the fundamental dimensionless constants of physics is altered by a few percent one way or the other. Man could never come into being in such a universe. That is the central point of the anthropic principle. According to the principle, a life-giving factor lies at the center of the whole machinery and design of the world."[11] Of course, thinking of man as the center of the universe has not had a strong track record in science, but let's set aside Copernican reservations in favor of contemporary astronomy.

Sir Martin Rees, Britain's Astronomer Royal, argues that "our emergence from a simple Big Bang was sensitive to six 'cosmic numbers.' Had these numbers not been 'well tuned,' the gradual unfolding of layer upon layer of complexity would have been quenched."[12] The six cosmic numbers are:

Ω (omega) = 1, the amount of matter in the universe, such that if Ω was greater than one, it would have collapsed long ago, and if Ω was less than one, no galaxies would have formed.

ε (epsilon) = .007, how firmly atomic nuclei bind together, such that if epsilon were .006 or .008, matter could not exist as it does.

D = 3, the number of dimensions in which we live, such that if D were 2 or 4, life could not exist.

N = 10^{36}, the ratio of the strength of gravity to that of electromagnetism, such that if it had just a few less zeros, the universe would be too young and too small for life to evolve.

Q = 1/100,000, the fabric of the universe, such that if Q were smaller, the universe would be featureless, and if Q were larger, the universe would be dominated by giant black holes.

λ (lambda) = 0.7, the cosmological constant, or "antigravity" force, that is causing the universe to expand at an accelerating rate, such that if λ were larger, it would have prevented stars and galaxies from forming.

Change these relationships and stars, planets, and life could not exist. Thus, this is not just the best of all possible worlds, it is the *only* possible world—and a world crafted with remarkable math skills, to boot. Intelligent Design theorists consider these numbers to be *complex* and *specified,* and thus the fine-tuned anthropic principle is evidence of design.[13]

First, the universe is not so finely tuned for life. The vast majority of the universe is empty space, and the vast majority of what little matter there is, is completely inhospitable to life, including most planets. In its 13.7-billion-year history, the anthropic conditions for life were nonexistent for several billion years—it is only during a narrow slice of recent time that the universe became finely tuned for life, and only a minuscule portion of the universe is hospitable. John Barrow and his colleague John Webb also note that the so-called "constants" of nature—the speed of light, gravitation, the mass of the electron—may be inconstant, varying over time from the Big Bang to the present, making the universe finely tuned only *now*.[14]

Second, our universe is not finely tuned for us (the strong anthropic principle), we are finely tuned for it (the weak anthropic principle). It is entirely possible that a completely different form of life could be based on another type of physics. We are carbon chauvinists, Carl Sagan liked to point out; life based on some other element (such as silicon) is entirely possible, but because we know of only one type, it is difficult for us to think outside the chemical box.

Third, our universe may not be that exceptional. String theory, for example, allows for 10^{500} possible worlds, all with different self-consistent laws and constants.[15] That's a 1-followed-by-500-zeros possible universes (recall that twelve zeros is a trillion!). If that is true, it would be miraculous if there were not intelligent life in a number of them. The physicist and astronomer Victor Stenger created a computer model that analyzes what just a hundred different universes would be like under constants different from our own, ranging from five orders of magnitude above to five orders of magnitude below their values in our universe. He discovered that long-lived stars of at least one billion years—necessary for the production of life-giving heavy elements—would emerge within a wide range of parameters in at least half of the universes in his model.[16]

Fourth, there may be an underlying principle behind all the fine-tune equations and relationships that will be forthcoming when the grand unified theory of physics is discovered. In the grand unified theory there will not be six mysterious numbers, there will be just one. Here we would do well to remember skeptical principle number two: *Before you say something is out of this world, first make sure it is not in this world.* Until we have a unified theory of physics connecting the quantum world of subatomic particles to the cosmic world of general relativity, we cannot conclude that there is something beyond nature to explain the anthropic principle.

Fifth, we may live in a *multiverse,* in which our universe is just one of many bubble universes, all with different laws of nature. Those with physical parameters like ours are more likely to generate life. Cosmologists theorize that there may even be a type of natural selection at work among the many bubble universes, in which those whose parameters are like ours are more likely to survive. According to inflationary cosmology, each time a black hole collapses, it does so into a singularity—the same entity out of which our universe may have sprung. Every time a star collapses into a black hole in our universe, the "other side" of the black hole may yield a new baby universe. Since there have likely been billions of collapsed black holes, there could be billions of bubble universes. Those universes whose initial conditions and physical laws do not produce stars like ours will not have black holes and thus will not reproduce more life-giving universes. Those bubble universes whose parameters are like ours are more likely to give rise to universes with life, perhaps even complex life with brains big enough to conceive of God and evolution.[17] How elegantly recursive!

In a slightly different scenario—one in which the universe is created out of a fluctuation in the quantum foam of space (it turns out that space is not so empty at the quantum level and that pure energy may give rise to matter)—Stephen Hawking answered the anthropic principle problem by conjecturing that new baby universes may be created in the same manner: "Quantum fluctuations lead to the spontaneous creation of tiny universes, out of nothing. Most of the universes collapse to nothing, but a few that reach a critical size, will expand in an inflationary manner, and will form galaxies and stars, and maybe beings like us."[18] Indeed, the multiverse is the next natural step in our expanding knowledge of the cosmos: from the earth to the solar system to the galaxy to the universe to the multiverse; that is, from the Copernican revolution that overturned the medieval worldview with the earth at the center and the stars and planets rotating close by on their crystal spheres and created within the last ten thousand years, to the early-modern worldview of the Milky Way galaxy as the entire known universe created within the last several million years, to the modern worldview of an accelerating expanding universe of some 13.7 billion years of age, to a multiverse of perhaps infinite age and containing perhaps an infinite number of universes.

Finally, from what we now know about the cosmos, to think that all this was created for just one species among tens of millions of species who live on one planet circling one of a couple of hundred billion stars that are located in one galaxy among hundreds of billions of galaxies, all of which are in one universe among perhaps an infinite number of universes all nestled within a grand cosmic multiverse, is provincially insular and anthropocentrically blinkered. Which is more likely? That the universe was designed just for us, or that we *see* the universe as having been designed just for us?

The Design Inference: There is a distinct difference between objects that are naturally designed and those that are intelligently designed.

Mount Rushmore is made entirely of natural material (rock), but no one would infer that the natural forces of erosion account for the design of four U.S. presidents' faces on the granite. This is an example of what Intelligent Design theorists call a "design inference," another staple argument, this one with its roots in lutemakers and William Paley's watchmakers. Of course, there are lots of examples of natural forces that do account for designed-looking objects: the rock formation in Maui's Iao Valley State Park that bears a striking resemblance to President John F. Kennedy in profile; the eroded mountain on Mars that under coarse-grained resolution looks like a face; the eagle rock off the 134 freeway in Southern California that overlooks the town of Eaglerock; the "Nun Bun" found by a Tennessee baker that resembles Mother Teresa; the Virgin Mary stained on the side of a bank building in Clearwater, Florida, or on a Chicago freeway underpass, or on a cheese sandwich in a Las Vegas casino. Although they were created entirely by natural forces, almost no one infers that there is an Intelligent Designer behind such artifacts of nature (with the possible exception of the Virgin Mary stains, which some religious devotees regard as miraculous apparitions). How can we tell the difference between natural design and artificial design?

"Design theorists infer a prior intelligent cause based upon present knowledge of cause-and-effect relationships," writes the philosopher of science and Intelligent Design advocate Stephen Meyer. "Inferences to design thus employ the standard uniformitarian method of reasoning used in all historical sciences, many of which routinely detect intelligent causes." Archaeologists, for example, employ statistical and physical criteria to discriminate between natural-made and human-made artifacts, so it is fair to say that "intelligent agents have unique causal powers that nature does not. When we observe effects that we know only agents can produce, we rightly infer the presence of a prior intelligence even if we did not observe the action of the particular agent responsible."[19] Intelligent Design theorists point to the elegance, uniformity, and ingenuity of DNA: It is no more naturally designed than the pyramids. If it looks intelligently designed, it was.

But the inference to design is subjective. Sometimes it is obvious, other times it is not. The difference between a rock and a watch is obvious; the difference between a rock and a chipped-stone tool made by an Australopithecene three million years ago is not obvious. And the inference to design is specific to each claim. In the chipped-stone problem, for example, a rock that has been chipped on both sides in a symmetrical fashion is more likely to be intelligently designed than naturally flaked. Nevertheless, archaeologists admit that they likely infer false positives, and there is no surefire design inference algorithm that applies to all archaeological problems, let alone one that applies to all scientific fields. The set of criteria used by archaeologists to determine whether a stone was chipped by chance or design is completely different from the set of criteria used by astronomers to determine whether a signal from space is natural or artificial.

Second, we perceive nature to be intelligently designed based on our experience of human artifacts. We know some human artifacts are intelligently designed because we have observed them being made and we have vast experience with human artificers. By contrast, we have no experience with a nonhuman intelligent designer, and no experience with a supernatural agent—outside of inferring that one exists by identifying the current gaps in our knowledge of apparently designed objects. The skeptical principle, *Methodological Naturalism, or, no miracles allowed,* refutes the inference to supernatural intelligent design. Not yet understanding how something was created naturally does not make it a supernatural creation.

Last, we must be cautious about inferring design because our experience with intelligently designed artifacts in our culture biases us to see intelligent design where none exists (for example, those Virgin Mary apparitions). Long before Darwin debunked the watchmaker argument, the Enlightenment philosopher Voltaire satirized this problem in his classic novel *Candide*, through the character Dr. Pangloss, a professor of "metaphysico-theology-cosmolonigology": "'Tis demonstrated that things cannot be otherwise; for, since everything is made for an end, everything is necessarily for the best end. Observe that noses were made to wear spectacles; and so we have spectacles. Legs were visibly instituted to be breeched, and we have breeches."[20]

Explanatory Filter: A tool for discriminating between natural design and intelligent design shows that only an Intelligent Designer can account for complex specified information and design.

Mathematician William Dembski has devised an "Explanatory Filter" through which intelligent design can be distinguished from natural design. Dembski asks, "When called to explain an event, object, or structure, we have a decision to make—are we going to attribute it to *necessity, chance,* or *design?*"[21] If necessity (natural law) and chance (randomness) cannot explain the phenomenon, then design (intelligence) is the default answer. The filter operates in a three-step process:

1. *Does natural law explain the design?* If event E has a high probability, accept *necessity* as an explanation; otherwise move to the next step.

2. *Does chance explain the design?* If event E has an intermediate probability or E is not specified, then accept *chance;* otherwise move to the next step.

3. *Does intelligent design explain the design?* Having eliminated necessity and chance as the explanation for a highly specified but low probability event, accept *design.*[22]

The Explanatory Filter is a tool by which we can tell the difference between the naturally designed JFK rock formation and the intelligently designed Mount Rushmore rock formation: "I argue that the Explanatory Filter is a reliable criterion for detecting design," Dembski explains. "Alternatively, I argue that the Explanatory Filter successfully avoids false positives. Thus, whenever the Explanatory Filter attributes design, it does so correctly."[23]

But the Explanatory Filter assumes probabilities that cannot be determined in practice; it is nothing more than a thought experiment and cannot practically be used in science. Indeed, in order to eliminate all necessity and chance explanations assumes that we know all configurations of these explanations, which, of course, we do not. Even if we did, and rejected them all, the design inference does not follow. Design, as it is commonly defined by the Intelligent Design movement, means *purposeful* and *intelligent* creation, not simply the elimination of necessity and chance. In other words, design is not just a default conclusion when all else fails to explain. Design requires positive evidence, not the rejection of negative evidence. The Explanatory Filter cannot survive the scrutiny of two skeptical principles—the *Burden of Proof* and the *Either-Or Fallacy*.

In addition, even if positive evidence for design were presented, by the logic of the Explanatory Filter we should be able to apply the filter to the design's designer. Assuming necessity and chance are mathematically (if not morally) rejected for the design's designer, the logical conclusion would be that the design's designer was designed, and that that design's designer's designer was also designed, ad infinitum. If the Explanatory Filter's design does not explain Intelligent Design, then by the logic of the Explanatory Filter we have to infer that necessity or chance created the Intelligent

Designer. By the logic of Shermer's Last Law, as we saw in the preceding chapter, the Explanatory Filter asserts that the Intelligent Designer and Extra-Terrestrial Intelligence are indistinguishable.

The ultimate answer to the design inference is to provide a cogent theory of natural design that can account for the complexity of the universe and life. This we have through the sciences of complexity, in which we recognize the properties of *self-organization* and *emergence* that arise out of *complex adaptive systems*. *Self-organization* means that the system requires only an input of energy into it in order to generate an action, which comes from within the system itself. An *emergent property* is one that is more than the sum of its parts. *Complex adaptive systems* are those that grow and learn as they change, and they are *autocatalytic*, which means that they contain self-driving feedback loops. For example:

Water is a self-organized emergent property of a particular arrangement of hydrogen and oxygen molecules.

Consciousness is a self-organized emergent property of billions of neurons firing in patterns in the brain.

Language is a self-organized emergent property of thousands of words spoken in communication between language users.

Law is a self-organized emergent property of thousands of informal mores and restrictions that were codified over time into formal rules and regulations as societies grew in size and complexity.

Economy is a self-organized emergent property of millions of people pursuing their own self-interests without any awareness of the larger system in which they work.

Life is a self-organized emergent property of prebiotic chemicals; complex life is a self-organized emergent property of simple life, where simple prokaryote cells self-organized to become more complex eukaryote cells (many of the little organelles inside cells were once self-contained independent cells); multicellular life is a self-organized emergent property of single-celled life; and so on up the chain of complexity to colonies, social units, societies, consciousness, language, law, and economies.

Self-organization is itself an emergent property, and emergence is a form of self-organization. The system is self-repeating, and therefore there is no need to invoke an Intelligent Designer.

The design inference comes naturally. The reason people think that a Designer created the world is because it *looks* designed. I think we should quit tiptoeing around this inference and admit that life looks designed because it *was*: from the bottom up, by evolution. The reason design appears artifactual to us is because evolutionary design is based on functional adaptation. Form follows function, function follows design, and natural selection selects among those designs that are most functional; that is, they enable the organism to survive and reproduce. When we use words like "design," "form," and "function," it sounds purposeful because we are accustomed to using these terms to describe human action, which we equate with purpose and intelligence. But in fact, the science of complexity shows how design, form, and function are all derivatives of self-organized emergent complex systems.

Irreducible Complexity: Evolution cannot account for
the stepwise gradual increase in complex systems.

In the *Origin of Species* Darwin wrote: "If it could be demonstrated that any complex organ existed which could not possibly have been formed by numerous, successive, slight modifications, my theory would absolutely break down."[24] Creationists have been in search of Darwin's exception ever since. Lehigh University biochemist Michael Behe thinks he has found several examples of living properties that he says are "irreducibly complex":

> By *irreducibly complex* I mean a single system composed of several well-matched, interacting parts that contribute to the basic function, wherein the removal of any one of the parts causes the system to effectively cease functioning. An irreducibly complex system cannot be produced directly (that is, by continuously improving the initial function, which continues to work by the same mechanism) by slight, successive modifications of a precursor system, because any precursor to an irreducibly complex system that is missing a part is by definition nonfunctional. Since natural selection can only choose systems that are already working, then if a biological system cannot be produced gradually it would have to arise as an integrated unit, in one fell swoop, for natural selection to have anything to act on.[25]

The human eye is a favorite example among Intelligent Design creationists because of its irreducible complexity—take out any one part and it will not work. How could natural selection have created the human eye when none of the individual parts themselves have any adaptive significance? Or consider the bacteria flagellum, Behe's type specimen of irreducible complexity and intelligent design—the little tail that propels the cell is complex and composed of many parts, and the removal of any one of them would cause the system to cease working. The bacteria flagellum is not *like* a machine, it *is* a machine, and it has no antecedents in nature from which it could have evolved in a stepwise Darwinian manner.

Irreducible complexity leads to an inference of intelligent design, an inference that Behe immodestly claims "is so unambiguous and so significant that it must be ranked as one of the greatest achievements in the history of science." He equates it with the discoveries of "Newton and Einstein, Lavoisier and Schrodinger, Pasteur, and"—with gumption—"Darwin."[26] There is even an emotional appeal to the struggle of Intelligent Design theorists for recognition of their scientific searches beyond the facile attribution of design to Darwin. "It is a shock to us in the twentieth century to discover, from observations science has made, that the fundamental mechanisms of life cannot be ascribed to natural selection, and therefore were designed. But we must deal with our shock as best we can and go on," Behe pleads. "The theory of undirected evolution is dead, but the work of science continues."[27]

Yes, the work of science does continue, and scientists in Behe's field of biochemistry and microbiology have responded to his claims.

First, when Behe says that "any precursor to an irreducibly complex system that is missing a part is by definition nonfunctional," he is committing the fallacy of bait-and-switch logic, says the philosopher of science Robert Pennock. He is reasoning from something that is true "by definition" to something that is proved through empirical evidence. Every time someone finds an example in nature that is simpler than Behe said it could be, Behe redefines irreducible complexity to *that* simpler level of complexity.[28] In other words, irreducible complexity is what Behe says it is, depending on the example at hand.

Second, the evolutionary biologist Jerry Coyne, responding to Behe directly, identified a number of biochemical pathways that

Behe has claimed are impossible to explain without an Intelligent Designer but which, in fact, "have been rigged up with pieces co-opted from other pathways, duplicated genes, and early multifunctional enzymes." Behe, for example, claims that the blood clotting process could not have come about through gradual evolution. Coyne shows that in fact thrombin "is one of the key proteins in blood clotting, but also acts in cell division, and is related to the digestive enzyme trypsin."[29] In other words, thrombin evolved for one purpose and was later coopted for other purposes.

Third, Behe's irreducible complexity is a more sophisticated version of an argument made against Darwin in the nineteenth century known as *the problem of incipient stages*. Fully formed wings are obviously an excellent adaptation for flight that provide all sorts of advantages for animals who have them; but of what use is half a wing? For Darwinian gradualism to work, each successive stage of wing development would need to be functional, but stumpy little partial wings are not aerodynamically capable of flight. Darwin answered his critics thus:

> Although an organ may not have been originally formed for some special purpose, if it now serves for this end we are justified in saying that it is specially contrived for it. On the same principle, if a man were to make a machine for some special purpose, but were to use old wheels, springs, and pulleys, only slightly altered, the whole machine, with all its parts, might be said to be specially contrived for that purpose. Thus throughout nature almost every part of each living being has probably served, in a slightly modified condition, for diverse purposes, and has acted in the living machinery of many ancient and distinct specific forms.[30]

This solution is called *exaptation,* in which a feature that originally evolved for one purpose is coopted for a different purpose.[31] The incipient stages in wing evolution had uses other than for aerodynamic flight—half wings were not poorly developed wings, they were well-developed something elses—perhaps thermoregulating devices. The first feathers in the fossil record, for example, are hairlike and resemble the insulating down of modern bird chicks.[32] Since modern birds probably descended from bipedal therapod dinosaurs, wings with feathers could have been employed for regulating heat—holding them close to the body would retain heat, stretching them out would release heat.[33] In the Galápagos Islands I have seen flightless cormorants returning to shore after diving for sea food, upon which they stretch out their stubby little wings with desultory feathers to dry them out and collect heat from the sun. In this case, wings evolved *from* flight tools *to* thermoregulation devices. In evolution, structures can be adapted for one function and evolve into use for another function and may have multiple functions at one time.

Another function for wings and feathers recently discovered is as an aid to running—some modern birds flap their wings to gain traction when running up steep inclines, even enabling them to climb straight up a 90-degree vertical structure.[34] Yet another function for incipient wings on bipedal dinosaurs was grasping. The most famous transitional fossil in evolutionary history, *Archaeopteryx,* has wings whose surface area is large enough to support its body in flight, asymmetrical feathers capable of attaining lift, and a shoulder that allows enough flexibility for an adequate upstroke of the wing necessary for flight. Nevertheless, *Archaeopteryx* retains many dinosaur features, including a functional grasping hand, for which the "wing" was probably originally adapted, and only later exapted for flight.[35]

Similar reasoning explains the incipient stages in the evolution of eyes, the flagellum motor, and the other structures claimed by Intelligent Design advocates to be inexplicable through evolutionary theory. For the human eye, it is not true that it is irreducibly complex, where the removal of any part results in blindness. Any form of light detection is better than none—lots of people are visually impaired by a variety of different diseases and injuries, yet they are able to utilize their restricted visual capacity to some degree, and would certainly prefer this to blindness. And natural selection did not create the human eye out of a warehouse of used parts lying around with nothing to do, any more than Boeing created the 777 without the ten million halting jerks and starts beginning with the Wright brothers.

As for the bacterial flagellum, although it is a remarkable structure, it comes in many varieties of complexity and functionality. Bacteria in general may be subdivided into *eubacteria* and *archaebacteria*; the former are more complex and have more complicated flagella, while the latter are simpler and have correspondingly simpler flagella. Eubacterial flagella, consisting of a three-part motor, shaft, and propeller system, are observedly a more complicated version of archaebacterial flagella, which have a motor and a combined shaft-propeller system. To describe the three-part flagellum as being irreducibly complex is just plain wrong—it can be reduced to two parts—and disingenuous. Additionally, the eubacterial flagellum turns out to be one of a variety of ways that bacteria move about their environment.[36] For many types of bacteria, the primary function of the flagellum is secretion, not propulsion. For others, the flagellum is used for attaching to surfaces and other cells.[37]

As to whether the flagellum is a product of evolution, eubacterial and archaebacterial flagella share similar structures owing to similar ancestry (known as *homologies*); there are also homologies between flagellar proteins and other systems, and functional similarity between the secretory systems and the propulsion systems of both eubacterial and archaebacterial flagella. Such similarities point to evolutionary development. We also know that eighteen to twenty genes are involved in the development of the simpler two-part flagellum, twenty-seven genes make up the slightly more complex *Campylobacter jejuni* flagellum, and forty-four genes exist in the still more complicated *E. coli* flagellum—a smooth genetic rise in complexity corresponding to the complexity of the end product. Finally, phylogenetic studies on flagella indicate that the more modern and complex systems share common ancestors with the simpler forms.[38] So here an evolutionary scenario presents itself: archaebacteria flagella were primarily used for secretion, although some forms were exapted for adhesion or propulsion. With the evolution of more complicated eubacteria, flagella grew more complex, refining, for example, the two-part motor and shaft-propeller system into a three-part motor, shaft, and propeller system that was then exapted for more efficient propulsion. Complex science reduced to evidence for evolution!

The Conservation of Information: Evolution cannot increase specific information content and complexity of organisms, or, there is No Free Lunch.

One of the most scientifically ambitious claims of Intelligent Design is William Dembski's *Law of Conservation of Information* (LCI), which is related to his analysis of *complex specified information* (CSI), an implicit ingredient in the four arguments above. LCI states that "natural causes are incapable of generating CSI" and that "the CSI in a closed system of natural causes remains constant or decreases." Dembski sets an upper limit to specified complexity generated by law and chance, which he calls the Universal Probability Bound (UPB), which he sets at $UPB = 1/2 \times 10^{-150}$, or about 500 bits of information. If the probability of a specified event is less than the UPB—that is, if it is over 500 bits of information—then it cannot be attributed to law or chance. By the logic of the Explanatory Filter, intelligent design is the best inference.[39]

This is quite a brainful, so restated, the Law of Conservation of Information says that natural causes such as chance and evolution cannot increase the complex specified information content of an organism beyond 500 bits of information. Since almost all living organisms contain more information than the Universal Probability Bound allows, without the insertion of complex specified information into living forms by an Intelligent Designer, the evolution of complex life is impossible.

Dembski's principles are embedded in a larger series of what he calls the "No Free Lunch theorems"—as though evolutionary theory were an attempt to grab a perk from the universe without paying for it. "The No Free Lunch theorems show that evolutionary algorithms, apart from careful fine-tuning by a programmer, are no better than blind search and thus no better than pure chance," and "the No Free Lunch theorems show that for evolutionary algorithms to output CSI they had first to receive a prior input of CSI" from an Intelligent Designer.[40] So confident is Dembski in his Law of Conservation of Information that he proposes it as a candidate for the Fourth Law of Thermodynamics.[41]

Dembski's Law of Conservation of Information is purposely constructed to resemble such physical laws as the conservation of momentum or the laws of thermodynamics. But these laws were based on copious empirical data and experimental results from the real world, not inferred from logical argument alone as Dembski's law is. Further, no other recognized theory of information—such as that proposed by the mathematician and information pioneer Claude Shannon—includes a law or principle of conservation, and no one working in the information sciences today uses or recognizes

Dembski's law as scientifically useful, regardless of its design inference.

Even if Dembski's LCI were validated, it is irrelevant to the theory of evolution, because it is abundantly clear that information in the natural world—through DNA, for example—is transferred and increased by natural processes. Microbiologist Lynn Margulis, for example, has demonstrated that complex eukaryote cells, such as those of which we are made, evolved from simpler prokaryote cells. The genomes of eukaryote cells increased in size—and thus in complex specified information—by incorporating simpler genomes of prokaryote cells.[42] Intelligent Design theorists respond to this by saying the Intelligent Designer created complex eukaryote cells. If that is so, why does it appear that the Intelligent Designer cobbled these cells together out of parts lying around in the pre-Cambrian soup? In fact, complex eukaryote cells are a grab bag of goodies from the prokaryote world, including mitochondria, which contain their own genome. Ever heard of mitochondrial DNA (from which human lineages may be traced through females)? Our cells already have a nucleus containing a complete genome. What is another genome doing in our mitochondria? Evolution offers an answer: They are vestigial features of eukaryote cells that evolved from prokaryote cells. Intelligent Design, in contrast, offers nothing more than "then a miracle occurs."

Richard Dawkins cogently answers the information challenge by reconstructing the evolution of hemoglobin—the oxygen-carrying protein in blood. Human hemoglobin contains four protein chains called *globins* that are similar to each other but not identical. The alpha globins each contain a chain of 141 amino acids coded by seven genes on Chromosome 11—four are pseudogenes that do not produce proteins, two produce adult hemoglobin, and one produces embryo hemoglobin. Similarly, beta globins each contain a chain of

146 amino acids coded by six genes on Chromosome 16, some of which are disabled and one used only in embryos. Letter-by-letter analysis of the genes coding for hemoglobin reveals that the two sets of genes on Chromosomes 11 and 16 are distantly related and share a common globin gene from a common ancestor five hundred million years ago. That gene duplicated, after which both copies were passed down for half a billion years, one evolving into the alpha cluster on Chromosome 11 and the other evolving into the beta cluster on Chromosome 16. Gene duplications led to the increase in complexity of the gene clusters, leading to the existence today of nonfunctional pseudogenes. Since this alpha-beta split happened five hundred million years ago we can predict that we should find the same alpha-beta split in all animals that evolved within the last half billion years. Sure enough, that is precisely what we find. As a final test, the jawless lamprey fish, the only surviving vertebrate predating the alpha-beta split, should lack this genetic divide. And sure enough, it does. Blood hemoglobin is explained by evolution, not Intelligent Design. Q.E.D.[43]

In general, DNA has the elements of historical contingency and evolutionary history, not design. DNA is information, and if the Law of Conservation of Information requires the input of an Intelligent Designer in order to increase specified complexity of the genome, we have to wonder why the Intelligent Designer added to our genome junk DNA, repeated copies of useless DNA, orphan genes, gene fragments, tandem repeats, and pseudogenes, none of which are involved directly in the making of a human being. In fact, of the entire human genome, it appears that only a tiny percentage is actively involved in useful protein production. Rather than being intelligently designed, the human genome looks more and more like a mosaic of mutations, fragment copies, borrowed sequences, and discarded strings of DNA that were jerry-built over millions of years of evolution.[44]

We Cannot Observe Evolution: No laboratory experiments or field observations reveal evolution in action.

Another regular on the Intelligent Design circuit is the hunt for "evolution in action." Scientists have not, and probably can never, provide examples of evolution at work that would satisfy an Intelligent Design creationist. It is one thing to infer in the fossil record the creation of a new anatomical structure, or the birth of a new species; it is quite another to witness it in the laboratory. And what examples we do have of evolution in action in the lab, creationists claim is not evolution.

But not only does science have an incredibly rich fossil record, the process of evolution can be seen at work at a number of different levels. Diseases are prime examples of natural selection and evolution at work, and on time scales we can witness, all too painfully. The AIDS virus, for example, continues to evolve in response to the drugs used to combat it—the few surviving strains of the virus continue to multiply, passing on their drug-resistant genes. Creationists respond that this is an example of microevolution, not macroevolution. Fair enough.

For an example of macroevolution, then, check out the research by the University of Michigan biologist James Bardwell, reported in the February 20, 2004, edition of *Science,* in which an *E. coli* bacterium that was forced to adapt or perish improvised a novel molecular tool. "The bacteria reached for a tool that they had, and made it do something it doesn't normally do. We caught evolution in the act of making a big step." The big step was a new way of making molecular bolts called *disulfide bonds,* which are stiffening struts in proteins that also help the proteins fold into their proper, functional, three-dimensional shapes. This new method restarted the bacterium's motor and enabled it to move toward food before it starved to death.

It is a particularly important experiment because Bardwell developed a strain of mutant bacteria unable to make disulfide bonds, which are critical for the ability of a bacterium's flagellum to work—the same flagellum that Intelligent Design theorists are so fond of presenting as an example of irreducible complexity. The researchers put these nonswimming bacteria to the test by placing them on a dish of food where, once they had exhausted the food they could reach, they either had to repair the broken motor or starve. The bacteria used in the experiment were forced to use a protein called *thioredoxin,* which normally destroys disulfide bonds, to make the bonds instead. In a process similar to natural selection, one researcher made random alterations in the DNA encoding thioredoxin and then subjected thousands of bacteria to the swim-or-starve test. He wanted to see if an altered version of thioredoxin could be coerced to make disulfides for other proteins in the bacteria. Remarkably, a mutant carrying only two amino acid changes—amounting to less than 2 percent of the total number of amino acids in thioredoxin—restored the ability of the bacteria to move. The altered thioredoxin was able to carry out disulfide bond formation in numerous other bacterial proteins all by itself, without relying on any of the components of the natural disulfide bond pathway. The mutant bacteria managed to solve the problem in time, swim away from starvation, and multiply.

Of course, Intelligent Design theorists will respond that the researcher acted as an intelligent designer would have in nature, and thus this supports their case; but, in fact, the researcher was acting as the force of natural selection, and thus this is evidence for evolution in action. Bardwell concluded that "the naturally occurring enzymes involved in disulfide bond formation are a biological pathway whose main features are the same from bacteria to man. People often speak of Computer Assisted Design (CAD), where you try things out on a computer screen before you manufacture them. We put the bacteria we were working on under a strong genetic selection, like what can happen in evolution, and the bacteria came up with a completely new answer to the problem of how to form disulfide bonds. I think we can now talk about Genetic Assisted Design (GAD)."

Perhaps we should now talk about GAD instead of GOD.

Microevolution and Macroevolution: Life shows signs of intermittent intelligent design intervention that accounts for large-scale changes.

Ever since Darwin, creationists have argued that natural selection can account for minor changes within a species, but cannot produce new species, new body forms, or new lineages. The argument presented today is a more sophisticated version in which, according to some (but not all) theorists, several billion years ago an Intelligent Designer created the first cell with the necessary genetic information to produce all of the irreducibly complex systems we see today. Then the laws of nature and evolution took over to create diversity within each species. When totally new and more complex species, body forms, and lineages appear in the fossil record, these are signs of the Intelligent Designer stepping in, intervening with a new design element. Microevolution proceeds by natural selection, but macroevolution is in the hands of the Designer.

First, how does one distinguish the processes of microevolution (evolution within and below the species level) from those of macroevolution (evolution above the species level)? Within evolutionary biology there has been considerable debate about whether the microevolutionary process of natural selection operating on individuals within populations can by itself account for the diverse macroevolutionary forms of life. Today, the new science of evolutionary developmental biology, "evo-devo" for short, reveals that the wide diversity of forms evolved through an interaction of the embryological development of forms and the subsequent pruning of these forms by natural selection.

For example, it turns out that the bodily architecture of vertebrates is the product of blueprint *Hox* genes that direct the construction of repeating parts, such as ribs and vertebrae. In embryological development, various structures form or do not form depending on whether the *Hox* genes are expressed or not. Natural selection operates on expressed forms only, since these result in organisms that survive long enough to pass on their genes for the future expression of those forms. Similarly, the wide variety of eyes found throughout the animal kingdom—from the compound eyes of flies to the camera eyes of vertebrates—evolved under the control of the commonly shared *Pax-6* gene, which directs the production of photoreceptor cells and light-sensing proteins. Each type of complex eye we find today evolved from simpler photoreceptive structures in a distant common ancestor of arthropods, cephalopods, and vertebrates. Evo-devo biologist Sean Carroll explains that

> the ancestor possessed two kinds of light-sensitive organs, each one endowed with a distinct type of photoreceptor, as well as with light-sensitive proteins called R-opsin and C-opsin, respectively. One organ was a simple two-celled prototype eye; the other, called the brain photoclock, was a part of the animal's brain and played a role in running the animal's daily clock. The arthropod and squid retinas incorporated the photoreceptor from the simple prototype eye, whereas the vertebrate eye incorporated both kinds of photoreceptor into its retina.[45]

Instead of eyes evolving forty or more different times in evolutionary history, it appears that this simple genetic complex led to the embryological development and evolutionary refinement of a two-part system; in some species one part is incorporated, and in others both are.

More generally, instead of an extensive genetic tool kit with genes for constructing each and every bodily structure, evo-devo shows that a small set of gene complexes such as the *Hox* genes and the *Pax-6* genes are expressed in novel ways that can generate large-scale changes in a nonincremental fashion. This explains why the human genome is not especially different from the mouse genome. It is not the number of genes that counts so much as how genes are turned on or off. Evolution involves old genes developing new tricks.

Second, a species is a group of actually or potentially interbreeding natural populations reproductively isolated from other such populations. We see evolution at work in nature today, isolating populations and creating new species, that is, new populations reproductively isolated from other such populations. As the new isolated populations drift genetically away from the parent populations, they eventually can no longer interbreed, making new species.[46] If evolution can do this, why can't it also create higher-order categories of organisms?

Third, some speciation may be precipitated by characteristics adapted to distinct environments that then drive populations into reproductive isolation, which leads to the creation of a new species. Similarly, sexual selection—female mate selection of males—may drive populations to diverge into different species. If females prefer certain traits in males, such as coloration, within one population, the males can change so dramatically that they are no longer appealing to females of another population, thereby making the two populations reproductively isolated: thus a new species. Research shows that speciation occurs more often in polygamous species than in monogamous species, further evidence linking sexual selection to the origin of new species.[47]

Fourth, to turn the tables on Intelligent Design theorists, how does Intelligent Design explain micro and macro forms? Did the Intelligent Designer personally tinker with the DNA of every single organism in a population? Or did the ID simply tweak the DNA of just one organism and then isolate that organism to start a new population? When and where did the ID intervene in the history of life? Did the ID create each genus and evolution then create each species? Or did the ID create each species and evolution create each subspecies? Most Intelligent Design theorists accept natural selection as a viable explanation for microevolution—the beak of the finch, the neck of the giraffe, the varieties of subspecies found on earth. If natural selection can create subspecies, why not species, genera, families, and on up the classification scale to kingdoms?

Last, just because Intelligent Design theorists cannot think of how nature could have created something through evolution, that does not mean that scientists will not be able to do so either. Intelligent Design is a remarkably uncreative theory that abandons the search for understanding at the very point where it is most needed. If Intelligent Design is really a science, then the burden is on its scientists to discover the mechanisms used by the Intelligent Designer. And if those mechanisms turn out to be natural forces, then no supernatural force is necessary, and they can simply change their name to evolutionary scientists and get to work.

*The Second Law of Thermodynamics makes evolution
impossible.*

According to the laws of physics, entropy increases—systems
change from hot to cold, from ordered to disordered, and from complex to simple. Yet evolutionists state that the universe and life
move from chaos to order and from simple to complex, the exact
opposite of the entropy predicted by the Second Law of Thermodynamics. Creationist Henry Morris stated the argument thus: "Evolutionists have fostered the strange belief that everything is involved
in a process of progress, from chaotic particles billions of years ago
all the way up to complex people today. The fact is, the most certain
laws of science state that the real processes of nature do not make
things go uphill, but downhill. Evolution is impossible."[48]

Yet on any scale other than the grandest of all—the three-billion-
year history of life on earth—species do not evolve from simple to
complex, and nature does not simply move from chaos to order.
The history of life is checkered with false starts, failed experiments, small and mass extinctions, and chaotic restarts. It is anything but the textbook foldout of linear progress from single cells to
humans.

Further, the Second Law of Thermodynamics applies to closed,
isolated systems. Since the earth receives a constant input of energy from the sun, it is an open-dissipative system, and entropy
may decrease and order increase (though the sun itself is running
down in the process). The earth is not strictly a closed system, and
life can evolve without violating natural law. As long as the sun is
burning, life may continue thriving and evolving, just as automo-
biles may be prevented from rusting, burgers can be heated in
ovens, and all manner of things in apparent violation of Second
Law entropy may continue. As soon as the sun burns out, entropy
will take its course—and life on earth will cease.[49]

In addition, an open-dissipative system such as we find on the earth slips in and out of thermodynamic equilibrium. The sciences of nonlinear dynamics and of chaos and complexity theory show that systems can spontaneously self-organize into more complex systems when they are in states of thermodynamic nonequilibrium. When a system is out of balance, energy flowing in and out of the system triggers the parts of the system to interact with one another locally, and these coupled interactions reverberate throughout the system to sustain it. Autocatalysis, or feedback loops within the system, can cause it to grow in complexity. From these self-organized autocatalytic interactions emerge complexity and order.[50] All of this happens without any top-down input. Evolution no more breaks the Second Law of Thermodynamics than one breaks the law of gravity by leaping into the air.

Evolution is random, and randomness cannot produce complex specified design.

It seems obvious that even the simplest of life forms are too complex to have come together by random chance. Take a simple organism consisting of merely 100 (10^2) parts. Mathematically there are 10^{158} possible ways for the parts to link up. There are not enough molecules in the universe, or time since the Big Bang, to account for these possible ways to come together in even this simple life form, let alone to produce human beings. It is the equivalent of a monkey randomly typing *Hamlet,* or even just the sentence "To be or not to be." It cannot happen by chance.

An understanding of evolutionary theory, however, makes clear that natural selection is not "random," nor does it operate by "chance." Natural selection preserves the gains and eradicates the mistakes. The eye evolved from a single light-sensitive cell into the modern complex eye through thousands of intermediate steps, many of which still exist in nature. Yes, in order for the monkey to type the first thirteen letters of Hamlet's soliloquy by chance, it would take 26^{13} trials (approximately 10^{18} times 2) to guarantee success—a number sixteen times as great as the total number of seconds that have elapsed in the lifetime of the solar system. But if each correct letter is preserved and each incorrect letter eradicated, as happens in natural selection, the process operates much faster. How much faster? My friend and colleague Richard Hardison constructed a computer program in which letters were "selected" for or against, and it took an average of only 335.2 trials to produce the sequence of letters TOBEORNOTTOBE, which on his computer took less than ninety seconds. The entire play can be generated in about four and a half days![51]

The icons of evolution are fallacies, fakes, or frauds.

Creationists of yore used to bleat about how the history of evolutionary science is nothing more than a catalogue of mistaken theories and overthrown ideas. Nebraska Man, Piltdown Man, Calaveras Man, and *Hesperopithecus*—all once claimed by scientists to be proof of human evolution—have all been shown to be mistakes or frauds. Clearly science cannot be trusted.

Intelligent Design theorists have modernized this argument through Jonathan Wells's book *Icons of Evolution: Science or Myth? Why Much of What We Teach About Evolution Is Wrong.*[52] Wells identifies ten "icons" of evolution, presented regularly in textbooks, that he says are mistakes, myths, or frauds:

1. *The Miller-Urey Experiment,* which demonstrates the chemical origins of the building blocks of life.
2. *Darwin's Tree of Life,* which shows that all living organisms come from a common ancestor.

3. *Homology in Vertebrate Limbs,* such as the similar bone structure of the forelimb of humans, bats, and whales, as examples of common ancestry and descent with modification of basic body plans.

4. *Haeckel's Embryo Drawings,* or "ontogeny recapitulates phylogeny," showing how in their embryological development (ontogeny) organisms go through similar stages in evolutionary (phylogeny) development.

5. *Archaeopteryx,* one of the best transitional fossils between dinosaurs and modern birds that demonstrates the evolution of flight.

6. *Peppered Moths,* which demonstrate how natural selection works by eliminating solid-colored moths that stand out on mottled tree bark, allowing the camouflaged peppered moths to survive and reproduce.

7. *Darwin's Finches,* which serve as examples of how speciation happens in nature.

8. *Four-Winged Fruit Flies,* which demonstrate how a change in a single gene can produce dramatic morphological and physical changes.

9. *Fossil Horses and Directed Evolution,* as examples of well-documented evolutionary pathways from simple to complex.

10. *From Ape to Human,* demonstrating the "missing links" between apes and humans.

When these are presented as examples of evolution, Wells argues, "students and the public are being systematically misinformed about the evidence for evolution." None of them is true, says Wells, and by easily refuting them, Intelligent Design creationists hope to bring down the entire evolutionary edifice.

First, the old creationists' saw that mistakes in science are a sign of weakness is a gross misunderstanding of the nature of science, which is constantly building upon both the mistakes and the successes of the past. Science does not just change, it builds cumulatively on the past. Scientists make mistakes aplenty, and in fact this is how science progresses. The self-correcting feature of the scientific method is one of its most powerful assets. Hoaxes like Piltdown Man and honest mistakes like Nebraska Man, Calaveras Man, and *Hesperopithecus* are in time exposed. In fact, it was not creationists who exposed these errors, it was scientists who did so. Creationists simply read about the scientific exposés of these errors, and then duplicitously advanced them as their own.

Intelligent Design theorists do the same thing, trolling through scientific journals and books in search of exposures by scientists of errors by other scientists and claiming them as the results of Intelligent Design research. This is what happened in the case of Wells's "punctured" icons of the peppered moths and Haeckel's embryo drawings. In the 1950s, the British evolutionary biologist Bernard Kettlewell correlated the evolution of darker forms of certain moths in England with industrial pollution—darker moths were better camouflaged on darker trees—and noted the reverse with the lightening of trees after clean air acts were passed (lighter moths were better camouflaged on lighter trees). But as later scientists noted, peppered moths do not rest on trees; although Kettlewell's recording of the changing coloration of peppered moths was accurate, his placing of them on trees was staged for photographic purposes and he was later nailed for it by scientists. In the case of Haeckel's embryological drawings, they were revealed to be fraudulent first in a technical scientific paper in the prestigious journal *Nature,* and then publicly by Stephen Jay Gould in his popular *Natural History* column.[53]

Next: Make that *nine* icons of evolution. The first icon Wells lists deals with the chemical origins of life, which is not generally considered a part of evolutionary theory. In any case, the science of life's origin has come a long way since Miller and Urey's experiments in the 1950s, so it is doubly disingenuous to focus on these primitive incipient research protocols.

Further, despite the high improbability of finding fossils, we have now a reasonably rich record of the history of life from the earliest bacteria fossils of 3.8 billion years ago, to the simple single-celled bacteria 3.5 billion years ago, to fossil archaebacteria 2.9 billion years ago, to the first eukaryotic cells about 2.5 billion years ago, to multicellular eukaryotic fossils 1.7 billion years ago, to fossils of higher algae forms 1.2 billion years ago, to the Ediacaran fauna 650 million years ago, all of which led to the so-called Cambrian "explosion" of about 550 to 500 million years ago. In highlighting this "explosion" as lacking any ancestral history (and therefore claiming it as proof that an Intelligent Designer miraculously sparked the creation of life), Intelligent Design advocates conveniently ignore three billion years of evidence of life's gradual evolution.

For each of these icons, Intelligent Design theorists also fail to provide an alternative theory to account for the data. For example, homology in vertebrate limbs has a perfectly sound evolutionary explanation: The skeletal structure of humans, bats, and whales is similar because we have similar ancestry—evolution created one basic architectural design from which natural selection shaped the wide variety of types. But even if this is wrong, as Wells says, what is the Intelligent Design explanation for homologies? Why would an Intelligent Designer design a whale's flipper to have precisely the same set of bones as a human arm and a bat's wing? And why would the ID make a bat's wing different from a bird's wing and an insect's wing? Evolution's answer is that a bat is a mammal with a different evolutionary lineage than birds and insects. What answer do Wells and the other Intelligent Design advocates offer?

Finally, there is the underlying assumption that the theory of evolution is founded on these ten icons, and thus that debunking them refutes the theory at large. Not so. The theory of evolution is proved through the convergence of evidence from thousands of lines of inquiry from diverse fields of study quite apart from this list. The vast bulk of data supporting evolution dwarfs these ten examples.

✦

That's the best science the Intelligent Design movement has to offer—lots of miracles, a handful of equations, and ten straw examples set against thousands of compelling lines of inquiry. But as often as not, science is not under debate; it is under attack. For example, in my debate at the University of California, Irvine, the Young Earth creationist Kent Hovind announced as his opening statement, "I am here to win you over to Christ. And I'm here to win Michael Shermer over to Christ."

With that statement, Hovind lost the debate. He was not there to debate evolution versus creation or natural versus supernatural design. He was there to witness for the Lord. Everything he said from there on was irrelevant or wrong: Dogs come only from dogs. Variations do not lead to new species. Design implies a Designer. There is an afterlife. The Bible is literally true in everything it says. Humans used to live nine hundred years. There is no right and wrong without God. Noah's flood explains geological formations and species distribution. Dinosaurs and humans lived simultaneously, dinosaurs on the Ark were very young and small, and dinosaurs that were large ("behemoth" and "leviathan" in the Bible) drowned in the flood. Radiometric dating is unreliable. Jesus said the universe is young. The theory of evolution is a religion that leads to atheism, abortion, and communism. Evolutionists are liars. Scientists are arrogant (they call themselves "Brights"!). Creationists are not allowed to publish in scientific journals. Creationism is censored from public schools. Microevolution may be true, but macroevolution, organic evolution, stellar evolution, chemical evolution, and cosmic evolution are all lies perpetrated by the lying liars who worship at the faux religion of evolution. And, of course, Jesus died for our sins.

This is what the evolution-creation debate is really about—religion, not science—and Intelligent Design theorists should be rightly called Intelligent Design creationists to drive the point home. Science is what scientists do, and Intelligent Design Creationists are not doing science. They are doing religion. It is not coincidental that almost all Intelligent Design creationists are Christians. But I will grant them this: *Intelligent Design arguments are reasons to believe if you already believe.* If you are not a True Believer, if you are a skeptic or a Fence Sitter, creationism and Intelligent Design are untenable.

Can any good come out of such debates? I think so. Outside heretics can stimulate us to refine our arguments and improve our explanatory prose. As Isaac Asimov once observed in confronting what he called *exoheresies,* or outside challenges to the status quo (in this case the radical ideas of Immanuel Velikovsky):

> An exoheresy may cause scientists to bestir themselves for the purposes of reexamining the bases of their beliefs, even if only to gather firm and logical reasons for the rejection of the exoheresy—and that is good.[54]

Yet, there is a final question that would chill Asimov's blood: Will those who accept evolution let those who do not accept it determine what science is?

5

SCIENCE UNDER ATTACK

———————◆———————

> How do we look for a new law? First, we guess it. Don't laugh.
> That's really true. Then we compute the consequences of the
> guess to see what it implies. Then we compare those compu-
> tation results to nature—or to experiment, or to experience, or
> to observation—to see if it works. If it disagrees with experi-
> ment, it is wrong. In that simple statement is the key to sci-
> ence. It doesn't make any difference how beautiful your guess
> is, how smart you are, who made the guess, or what his name
> is. If it disagrees with experiment, it's wrong. That's all there is
> to it.
>
> —Richard Feynman, lecture at Cornell University
> on the nature of science, 1964

So much comes down to necessity and chance.

In the 1990s I published a series of articles in respected peer-
reviewed journals applying chaos and complexity theory to human
history.[1] Out of that research I constructed a theoretical model
demonstrating the relative roles of necessity and contingency—law
and chance—in history, and how the relationship of these two factors helps to explain why sometimes the kingdom is lost for the
want of a horseshoe nail while at other times a million horseshoe
nails would have made no difference at all. Sometimes "great men"
make history; at other times geographical conditions, social circumstances, economic forces, and political machinations swamp

any influence that individuals may have. I even published a book with one of the top university presses in which I demonstrated how my theory helps to explain such world-shaping events as the Holocaust.[2] I had high hopes that historians would adopt my theory, put it into practice, and perhaps even teach it to their students. They haven't. Maybe I did not communicate my theory clearly. Possibly historians do not use such theoretical models. Worse, perhaps my theory is wrong or useless. Should I appear before Congress to demand that legislation be passed to give my theory equal time with other theories of history? Should I lobby school board members to force history teachers to teach my theory of history?

The God of the Government

Since Intelligent Design theory has failed to win the debate on scientific merit, or to convince scientists to accept its ideas as providing some useful insight into evolution and the structure of life, many of its proponents are taking their case to the government. If they cannot get the scientists to believe in their ideas, they will legislate their ideas into the classroom. The reasoning is rather straightforward:

1. Scientists do not accept Intelligent Design as science.
2. Therefore Intelligent Design is not taught in public school science classes.
3. I think Intelligent Design is science.
4. Therefore I will lobby the government to force teachers to teach Intelligent Design as science.

This is what I call the *God of the Government* argument (*pace the God of the Gaps* argument discussed in the last chapter): *If you*

*can't persuade teachers to teach your idea based on its own merits, see
if you can get the government to force teachers to teach it.* If I were
trying to force my theory of history into public school history
classes, my actions would be considered ludicrous. It is just as absurd when Intelligent Design theorists push their way into science lesson plans. In the free marketplace of ideas, turning to
the government to force your theory on others—particularly
children—goes against every principle of liberty upon which modern Western democracies are founded. It seems, however, that science might not fall under such moral principles of liberty—unless
we fight for it.

If I were a religious believer, I would be embarrassed by the latest round of attempts to legislate these beliefs into the public
schools. If creationists want their doctrines taught in public school
science classes, they need first to develop a science, and then to
convince scientists that their scientific ideas merit inclusion based
on the quality of the arguments and evidence.

How Science Makes It into
Science Textbooks and Classrooms

Since we live in a free society, parents are free to choose whatever
schools they want their children to attend, or even to homeschool
their children if they are dissatisfied with the choice of public and
private schools in their area. If all schools were private, and if the
education of children were strictly a function of the free market,
there would be no high-profile court cases and school board battles
over evolution and creationism; there would be no debate over evolution and Intelligent Design. Creationist parents would be free to
send their kids to private schools where creationism is taught.

Indeed, some creationist parents do this now (or they opt for homeschooling programs that include a creationist unit in the biology curriculum).

Conflict arises out of the fact that public schools are funded by the government, and since *we* are the government, taxpayers feel that they should have some say in what is taught in public schools. This sounds like a reasonable argument, until we carry it to its logical conclusion—all parents would be justified in demanding equal time for their particular religious, political, or social beliefs. Christians would want a Christian slant in the curriculum, Muslims would want a Muslim slant, Native Americans a Native American slant, and so forth. Education would dissolve into an endless parade of beliefs given equal time, with no core curriculum on which to focus students' attention. (To get just a flavor of exactly what that "so forth" would entail, visit the appendix.)

So how does a new scientific discovery or theory make it into the science curriculum? It usually takes a long time, because science is a fairly conservative institution with high and exacting standards of evidence. It is typically years before experimental results trickle down from scientific conferences and journals into textbooks and lecture notes; and it is often decades before a new theory displaces an existing and commonly taught theory. Scientists face these hurdles all the time. The case of the microbiologist Lynn Margulis is instructive.

Lynn Margulis is best known for her theory of *symbiogenesis*, which challenges the tenet that inherited variation comes primarily from random mutations. Rather, Margulis argues, new species, at least microbial species, evolve through the exchange of genomes, where the fusion of genomes in symbioses leads to the variation on which natural selection acts, which then leads to increasing complexity in the species. She first published her theory in 1970. For more than three decades she has been lecturing on it at scientific conferences, writing about it in hundreds of articles in peer-reviewed scientific journals, expanding upon it in technical books published by peer-reviewed university presses, and elaborating on it in popular books published by trade presses. After all this effort and evidence, the theory of symbiogenesis is finally wending its way into the generally accepted body of evolutionary knowledge taught to students, even though it remains controversial in some scientific circles.[3] If Intelligent Design creationists want to know how to get their theory taught in public schools, they should take a lesson from Lynn Margulis: Roll up your sleeves and get to work—lab and field work, not legislative lobby work.

This vetting process by the community of scientists in a field is how new discoveries and theories gain acceptance or experience rejection. Adjudication is determined by a vote of sorts—the scientists in a field vote with their feet, either by running back to their labs to test the new discovery or theory, or by discarding it altogether. If it is useful, it stands a good chance of finding its way into general textbooks, which are often written by members of that scientific community. This system sounds insular, but it is remarkably egalitarian and democratic because anyone can join in the process, as long as they abide by the rules of the game of science. The Intelligent Design theorists, rather than respect these rules, instead revert to the question, *What is science?*

Science Defined in Its Defense

Creationists sometimes claim that the theory of evolution is a doctrine in a religion they call Secular Humanism, and thus if creationism is not taught in public schools, then neither should evolution be.

The vast majority of believers and theists around the world fully accept the theory of evolution, so clearly they are not mutually exclusive. But is the theory of evolution a religious belief? No, it is not.

If a branch of science like evolutionary theory is a tenet of religion, then the definition of religion is so sweeping that virtually everything is a religion, rendering the word meaningless. Science is not a religion. Science is a very specific branch of human knowledge with a set of methods quite distinct from other branches of knowledge. I have, in earlier chapters, pragmatically defined it this way: *Science is a testable body of knowledge open to rejection or confirmation.* More formally, I have defined it as follows:

> Science is a set of methods designed to describe and interpret observed or inferred phenomena, past or present, aimed at building a testable body of knowledge open to rejection or confirmation.[4]

The description of methods is essential, however, because it shows how science actually works. Included in the methods are hunches, guesses, ideas, hypotheses, theories, and paradigms, and testing them involves background research, experiments, data collection and organization, colleague collaboration and communication, correlation of findings, statistical analyses, conference presentations, and publications. In the simplest sense, science is what scientists do.

Although there is much debate among philosophers and historians of science about what science is, there is general agreement that science revolves around what is known formally as the *hypothetico-deductive method:* (1) formulating a hypothesis, (2) making a prediction based on the hypothesis, and (3) testing whether or not the prediction is accurate. In formulating hypotheses and theories, science employs natural explanations for natural phenomena.

These characteristics of science were even codified into law in two important evolution-creationism trials in the 1980s, one in Arkansas and the other in Louisiana, the latter of which was appealed up to the U.S. Supreme Court. When forced to do so, science has defined itself, in defense.

The 1981 Arkansas trial was over the constitutionality of the state's Act 590, which required equal time in public school science classes for "creation-science" and "evolution-science." The federal judge in that case, William R. Overton, ruled against the creationists on the following grounds: First, he said, creation-science conveys "an inescapable religiosity" and is therefore unconstitutional: "Every theologian who testified, including defense witnesses, expressed the opinion that the statement referred to a supernatural creation which was performed by God." Second, Overton said that the creationists employed a "two model approach" in a "contrived dualism" that "assumes only two explanations for the origins of life and existence of man, plants and animals: It was either the work of a creator or it was not." In this either-or paradigm, creationists claim that any evidence "which fails to support the theory of evolution is necessarily scientific evidence in support of creationism." Overton slapped down the tactic, writing "evolution does not presuppose the absence of a creator or God."

More important, Judge Overton summarized why creation-science is not science by explaining what science is:

1. It is guided by natural law.
2. It has to be explanatory by reference to natural law.
3. It is testable against the empirical world.
4. Its conclusions are tentative.
5. It is falsifiable.

Overton concluded: "Creation science as described in [the Act's] Section 4(a) fails to meet these essential characteristics," adding the "obvious implication" that "knowledge does not require the imprimatur of legislation in order to become science."[5]

The 1987 Louisiana case amplified the description of science even more because this case was appealed all the way to the U.S. Supreme Court, thereby fulfilling the ACLU's original intent for the 1925 Scopes Tennessee trial. For the case of *Edwards v. Aguillard*, seventy-two Nobel laureates, seventeen state academies of science, and seven other scientific organizations submitted an *amicus curiae* brief to the Court's justices in support of the appellees' challenge of the constitutionality of Louisiana's "Balanced Treatment for Creation-Science and Evolution-Science Act," an equal-time law passed by the state in 1982. The brief is one of the most important documents in the history of the evolution-creation debate and presents the best short statement on the central tenets of science endorsed by the world's leading scientists and science organizations.[6]

The brief responds to all of the attacks on evolution and science, opening with a demonstration that "creation-science" is just a new label for the old religious creationism of decades past. It then defines the criteria of science, a field "devoted to formulating and testing naturalistic explanations for natural phenomena. It is a process for systematically collecting and recording data about the physical world, then categorizing and studying the collected data in an effort to infer the principles of nature that best explain the observed phenomena." At the heart of science is the scientific method, and these preeminent scientists took their opportunity to enter it into the Court record. From facts to hypotheses to theories to conclusions to explanations, the toolkit of science became, for better or worse, a subject for government decision:

Facts. "The grist for the mill of scientific inquiry is an ever increasing body of observations that give information about underlying 'facts.' Facts are the properties of natural phenomena. The scientific method involves the rigorous, methodical testing of principles that might present a naturalistic explanation for those facts."

Hypotheses. Based on well-established facts, testable hypotheses are formed. The process of testing "leads scientists to accord a special dignity to those hypotheses that accumulate substantial observational or experimental support."

Theories. This "special dignity" is called a "theory" that, when it "explains a large and diverse body of facts," is considered "robust," and, if it "consistently predicts new phenomena that are subsequently observed," is deemed "reliable." Facts and theories are not to be used interchangeably or in relation to one another as more or less true. Facts are the world's data. Theories are explanatory ideas about those data. Constructs and other nontestable statements are not a part of science. "An explanatory principle that by its nature cannot be tested is outside the realm of science."

Conclusions. It follows from this process that no explanatory principles in science are final. "Even the most robust and reliable theory . . . is tentative. A scientific theory is forever subject to reexamination and—as in the case of Ptolemaic astronomy—may ultimately be rejected after centuries of viability. In an ideal world, every science course would include repeated reminders that each theory presented to explain our observations of the universe carries this qualification: 'as far as we know now, from examining the evidence available to us today.'"

Explanations. Science also seeks only naturalistic explanations for phenomena. "Science is not equipped to evaluate
supernatural explanations for our observations; without
passing judgment on the truth or falsity of supernatural explanations, science leaves their consideration to the domain
of religious faith." Any body of knowledge accumulated
within these guidelines is considered "scientific" and suit-
able for public school education; and any body of knowledge
not accumulated within these guidelines is not considered
scientific. "Because the scope of scientific inquiry is con-
sciously limited to the search for naturalistic principles,
science remains free of religious dogma and is thus an appropriate subject for public-school instruction."

This case was decided on June 19, 1987, with the Court voting
7–2 in favor of the appellees, holding that "the Act is facially in-
valid as violative of the Establishment Clause of the First Amend-
ment, because it lacks a clear secular purpose" and that "[t]he Act
impermissibly endorses religion by advancing the religious belief
that a supernatural being created humankind." Predictably, Justices Antonin Scalia and William Rehnquist dissented, arguing that
"so long as there was a genuine secular purpose," the Christian
fundamentalist intent "would not suffice to invalidate the Act." Re-
calling the academic freedom issue that was argued more than
sixty years before in the Scopes trial, Scalia and Rehnquist note,
"The people of Louisiana, including those who are Christian fun-
damentalists, are quite entitled, as a secular matter, to have whatever scientific evidence there may be against evolution presented
in their schools, just as Mr. Scopes was entitled to present whatever scientific evidence there was for it." The majority of the Court
disagreed, stating that regardless of the religious intent of the

creationists, there is no science in creation-science; there is strong evidence that their opinions were shaped by the *amicus curiae* brief that demonstrated so plainly why creationism is not science by clearly explaining what science is.

The ongoing court cases and curriculum battles being fought over Intelligent Design creationism involve the same issues that were settled in the 1987 U.S. Supreme Court decision. But Intelligent Design creationists are repackaging their wares, appealing to our human natures, and taking advantage of a larger attack on science. There is no more science in Intelligent Design theory than there is in creation-science; but the point of the movement is not to expand scientific understanding—it is to shut it down. Case in point: *Kitzmiller et al. v. Dover Area School District,* the first evolution-creationism trial of the twenty-first century.

Design in Dover

In the legendary debate over evolution at Oxford University in June 1860, Archbishop Samuel Wilberforce ("Soapy Sam") sardonically inquired of his debate opponent, Thomas Henry Huxley ("Darwin's Bulldog"), whether he was descended from an ape on his grandfather's or grandmother's side. Accounts vary as to what was said next, but legend has it that Huxley muttered to the person next to him, "The Lord hath delivered him into my hands," and followed that with his stinging rejoinder to Soapy Sam: "If then, said I, the question is put to me would I rather have a miserable ape for a grandfather or a man highly endowed by nature and possessed of great means of influence and yet who employs these faculties and that influence for the mere purpose of introducing ridicule into a grave scientific discussion, I unhesitatingly affirm my preference for the ape."[7]

This is precisely how I felt when the judge issued his decision in the *Kitzmiller* case in late 2005. It was as if the Lord had delivered the creationists into our hands, and the judge's devastating critique of the creationist defense would make even the staunchest religious conservatives unhesitatingly affirm their preference for evolution. *Kitzmiller* was an exceptional court case—both for what it revealed about the motives of the Intelligent Design creationists, and the clarity and severity of the conservative judge's decision against the Intelligent Design proponents.[8]

In the trial, the Dover Area School District was defended by the Thomas More Law Center (TMLC), an organization founded by conservative Catholic businessman Tom Monaghan and attorney Richard Thompson, who prosecuted Jack Kevorkian in the very public trial over assisted suicide. From its founding in 1999, the TMLC has been searching for venues in which to take on the ACLU. Calling themselves the "Christian Answer to the ACLU" and the "sword and shield for people of faith," the TMLC has challenged the ACLU on a range of public controversies, from pornography and gay marriage to nativity scenes and Ten Commandment displays. Starting in early 2000, representatives of the TMLC canvassed school boards around the country, searching for and encouraging the teaching of Intelligent Design in public school science classrooms. The TMLC recommended that biology teachers supplement their standard textbook with the textbook *Of Pandas and People*. That creationist tract would prove to be a key piece of evidence in the trial. In 2004 the TMLC found a willing accomplice in the school board for the Dover Area School District in Pennsylvania, which was dominated by conservative Christians seeking a way to introduce creationism into their children's science classrooms.

On October 18, 2004, the school board met and voted 6–3 to add the following statement to their biology curriculum: *"Students will be made aware of the gaps/problems in Darwin's theory and of other theories of evolution including, but not limited to, intelligent design. Note: Origins of life is not taught."* The next month the board added a statement to be read to all ninth-grade biology classes at Dover High:

> The Pennsylvania Academic Standards require students to learn about Darwin's theory of evolution and eventually to take a standardized test of which evolution is a part.
>
> Because Darwin's Theory is a theory, it is still being tested as new evidence is discovered. The Theory is not a fact. Gaps in the Theory exist for which there is no evidence. A theory is defined as a well-tested explanation that unifies a broad range of observations.
>
> Intelligent design is an explanation of the origin of life that differs from Darwin's view. The reference book, *Of Pandas and People,* is available for students to see if they would like to explore this view in an effort to gain an understanding of what intelligent design actually involves.
>
> As is true with any theory, students are encouraged to keep an open mind. The school leaves the discussion of the origins of life to individual students and their families. As a standards-driven district, class instruction focuses upon preparing students to achieve proficiency on standards-based assessments.

Copies of the book *Of Pandas and People* were made available to the school by William Buckingham, the chair of the curriculum committee, who raised $850 from his church to purchase copies of the book for the school. As he told a Fox-TV affiliate in an interview the week after the school board meeting, "My opinion, it's okay to teach Darwin, but you have to balance it with something else such as creationism." But eleven parents of students enrolled in Dover High would have none of this, and on December 14, 2004, they filed suit against the district with the legal backing of the ACLU and Americans United for Separation of Church and State. The TMLC had the fight they were aching for. The suit was brought in the U.S. District Court for the Middle District of Pennsylvania, and a bench trial was held from September 26 to November 4, 2005, presided over by Judge John E. Jones III, a conservative Christian appointed to the bench in 2002 by President Bush.

The primary task of the prosecution was to show not only that Intelligent Design is not science but that it is just another name for creationism, which the U.S. Supreme Court had already decided in *Edwards v. Aguillard*—the Louisiana case—could not be taught in public schools. Expert scientific witnesses testified on behalf of the prosecution, including Brown University molecular biologist Kenneth Miller and University of California, Berkeley, paleontologist Kevin Padian, both of whom rebutted specific Intelligent Design claims. More important were the expert testimonies of the philosophers Robert Pennock, from Michigan State University, and Barbara Forrest, from Southeastern Louisiana University, both of whom had authored definitive histories of the Intelligent Design movement. Pennock and Forrest presented overwhelming evidence that Intelligent Design is, in the memorable phrase of one observer, nothing more than "creationism in a cheap tuxedo."

It was revealed, for example, that the lead author of the book *Of Pandas and People*, Dean Kenyon, had also written the foreword to the classic creationism textbook *What Is Creation Science?* by Henry Morris and Gary Parker. The second author of *Pandas*, Percival Davis, was the co-author of a Young Earth creationism book called *A Case for Creation*. But the most damning evidence was in the book itself. Documents provided to the prosecution by the National Center for Science Education revealed that *Of Pandas and People* was originally titled *Creation Biology* when it was conceived in 1983, then *Biology and Creation* in a 1986 version, which was retitled yet again a year later to *Biology and Origins*. Since this was before the rise of the Intelligent Design movement in the early 1990s, the manuscripts referred to "creation," and fund-raising letters associated with the publishing project noted that it supported "creationism." The final version, by now titled *Of Pandas and People,* was released in 1989, with a revised edition published in 1993. Interestingly, in the 1986 draft, *Biology and Creation,* the authors presented this definition of the central theme of the book, creation, as follows:

> Creation means that the various forms of life began abruptly through the agency of an intelligent creator with their distinctive features already intact. Fish with fins and scales, birds with feathers, beaks, and wings, etc.

Yet, in *Of Pandas and People,* published after *Edwards v. Aguillard,* the definition of creation mutated to this:

> Intelligent design means that various forms of life began abruptly through an intelligent agency, with their distinctive features already intact. Fish with fins and scales, birds with feathers, beaks, wings, etc.

So there it was, the smoking gun. The textbook recommended to students as the definitive statement of Intelligent Design began its evolving life as a creationist tract. Like the old Monty Python routine where the guy changes a dog license to a cat license by simply crossing out "dog" and writing in "cat," the creationists simply deleted "creation" and pasted in "intelligent design."

If all this were not enough to indict the true motives of the creationists, the prosecution punctuated the point by highlighting a statement made by the purchaser of the school's copies of *Pandas,* William Buckingham, who told a local newspaper that the teaching of evolution should be balanced with the teaching of creationism because "[t]wo thousand years ago, someone died on a cross. Can't someone take a stand for him?"

This was all too much even for the ultra-conservative Judge Jones. On the morning of December 20, 2005, he released his decision—a ringing indictment of both Intelligent Design and religious insularity:

> The proper application of both the endorsement and Lemon tests to the facts of this case makes it abundantly clear that the Board's ID Policy violates the Establishment Clause. In making this determination, we have addressed the seminal question of whether ID is science. We have concluded that it is not, and moreover that ID cannot uncouple itself from its creationist, and thus religious, antecedents.

Judge Jones went even further, excoriating the board members for their insistence that evolutionary theory contradicts religious faith:

> Both Defendants and many of the leading proponents of ID make a bedrock assumption which is utterly false. Their presupposition is that evolutionary theory is antithetical to a belief in the existence of a supreme being and to religion in general. Repeatedly in this trial, Plaintiffs' scientific experts testified that the theory of evolution represents good science, is overwhelmingly accepted by the scientific community, and that it in no way conflicts with, nor does it deny, the existence of a divine creator.

Demonstrating his understanding of the provisional nature of science, Judge Jones added that uncertainties in science do not translate into evidence for a nonscientific belief:

To be sure, Darwin's theory of evolution is imperfect. However, the fact that a scientific theory cannot yet render an explanation on every point should not be used as a pretext to thrust an untestable alternative hypothesis grounded in religion into the science classroom or to misrepresent well-established scientific propositions.

The judge pulled no punches in his opinion about the board's actions and especially their motives, going so far as to call them liars:

The citizens of the Dover area were poorly served by the members of the Board who voted for the ID Policy. It is ironic that several of these individuals, who so staunchly and proudly touted their religious convictions in public, would time and again lie to cover their tracks and disguise the real purpose behind the ID Policy.

Finally, knowing how his decision would be received by the press, Judge Jones forestalled any accusations of him being an activist judge, and in the process took one more shot at the "breathtaking inanity" of the Dover school board:

Those who disagree with our holding will likely mark it as the product of an activist judge. If so, they will have erred as this is manifestly not an activist Court. Rather, this case came to us as the result of the activism of an ill-informed faction on a school board, aided by a national public interest law firm eager to find a constitutional test case on ID, who in combination drove the Board to adopt an imprudent and ultimately unconstitutional policy. The breathtaking inanity of the Board's decision is evident when considered against the factual backdrop which has now been fully revealed through this trial. The students, parents, and teachers of the Dover Area School District deserved better than to be dragged into this legal maelstrom, with its resulting utter waste of monetary and personal resources.

Q.E.D.

6

THE REAL AGENDA

———————— ✦ ————————

Johnson calls his movement "The Wedge." The objective, he said, is to convince people that Darwinism is inherently atheistic, thus shifting the debate from creationism vs. evolution to the existence of God vs. the non-existence of God. From there people are introduced to "the truth" of the Bible and then "the question of sin" and finally "introduced to Jesus."

—Rob Johnson, on ID proponent Phillip Johnson,
Church & State magazine, 1999

One evening several years ago, while on a book tour for *How We Believe,* I gave a lecture at MIT on why people believe in God. Coincidentally, at the same time, down the hall, the mathematician William Dembski was lecturing on Intelligent Design theory. After our respective talks we did what any two people of opposing camps in a controversy should do—we went out for a beer. Accompanied by Bill's colleague Paul Nelson, we sat around a sports bar and reflected on science and religion, evolution and creationism, and—this being Boston—the Red Sox and the Yankees. Since that evening, I have debated Bill, Paul, and the Intelligent Design philosopher Stephen Meyer, on several occasions, and have shared car rides and meals in the process. Paul Nelson visited the Skeptics Society office and library, after which we dined with God and mammon. Having gotten to know these gentlemen over the years, I must aver that a more gracious, considerate, and thoughtful group you will not find.

Because of our friendship, these guys have been forthright with me about their religious beliefs, which, of course, I could not help but inquire about. Although to a man they remain steadfast in their claim that they are pursuing a scientific agenda and not a religious one, they privately acknowledge their belief that the Intelligent Designer is the God of Abraham. To my knowledge, in fact, all but one of the leading Intelligent Design proponents is an evangelical Christian.

On the one hand, this should not matter in the assessment of someone's claim, and I have devoted the longest chapter of this book to their arguments. On the other hand, when nearly every single member of a scientific community belongs to one particular religious faith, your baloney detection alarms should signal you that there is something else afoot here, as indeed there is.

As a scientist, I look to the data. And although I disdain to accuse friends of being insincere about their motives, the extant evidence—in their own published words—leads me to conclude that there is a distinct and definite religious and political agenda behind and above whatever science they think they are pursuing. Human behavior is complex and multivariate in its causes—motives are not so easily pigeonholed into black-and-white categories. In my opinion, the Intelligent Design creationists I have met believe their own rhetoric about only doing science and having no religious or political agendas, and they also believe in the religious and political tenets to which they adhere.

God and the Wedge

In an attempt to distance themselves from "scientific creationists," who were handily defeated in the 1987 Supreme Court case, Intelligent Design creationists emphasize that they are interested only in doing science. According to Dembski, for example, "scientific creationism has prior religious commitments whereas intelligent design does not."[1]

Baloney. On February 6, 2000, Dembski told the National Religious Broadcasters at their annual conference in Anaheim, California, that "intelligent design opens the whole possibility of us being created in the image of a benevolent God. . . . The job of apologetics is to clear the ground, to clear obstacles that prevent people from coming to the knowledge of Christ. . . . And if there's anything that I think has blocked the growth of Christ as the free reign of the Spirit and people accepting the Scripture and Jesus Christ, it is the Darwinian naturalistic view."[2] In a feature article in the Christian magazine *Touchstone,* Dembski was even more direct: "Intelligent design is just the Logos theology of John's Gospel restated in the idiom of information theory."[3]

Make no mistake about it. Creationists and their Intelligent Design brethren do not just want equal time, they want all the time they can get. Listen to the words of Phillip Johnson, the University of California, Berkeley, law professor who is the fountainhead of the modern Intelligent Design movement, at the same National Religious Broadcasters meeting at which Dembski spoke: "Christians in the twentieth century have been playing defense. They've been fighting a defensive war to defend what they have, to defend as much of it as they can. It never turns the tide. What we're trying to do is something entirely different. We're trying to go into enemy territory, their very center, and blow up the ammunition dump. What is their ammunition dump in this metaphor? It is their version of creation."[4] In 1996, Johnson did not pull his punches: "This isn't really, and never has been, a debate about science. . . . It's about religion and philosophy."[5]

Enter the Wedge. It was Johnson who introduced the metaphor in his book *The Wedge of Truth*. "The Wedge of my title is an informal movement of like-minded thinkers in which I have taken a leading role," he writes. "Our strategy is to drive the thin end of our Wedge into the cracks in the log of naturalism by bringing long-neglected questions to the surface and introducing them to public debate." After naturalism falls, materialism is the next target in their gun sights. "Once our research and writing have had time to mature, and the public prepared for the reception of design theory, we will move toward direct confrontation with the advocates of materialist science through challenge conferences in significant academic settings. . . . The attention, publicity, and influence of design theory should draw scientific materialists into open debate with design theorists, and we will be ready."[6] This is not just an attack on naturalism, it is a religious war against all of science. "It is time to set out more fully how the Wedge program fits into the specific Christian gospel (as distinguished from generic theism), and how and where questions of biblical authority enter the picture. As Christians develop a more thorough understanding of these questions, they will begin to see more clearly how ordinary people—specifically people who are not scientists or professional scholars—can more effectively engage the secular world on behalf of the gospel."[7]

The principal exception to my earlier generalization that Intelligent Design creationists are Christians is the author of the Top Ten list of evolutionary "icons," Jonathan Wells. Wells is a Moonie—a member of the Unification Church and a follower of the Reverend Sun Myung Moon, who assigned Wells the task of destroying evolution. "Father's [the Reverend Sun Myung Moon's] words, my studies, and my prayers convinced me that I should devote my life to destroying Darwinism, just as many of my fellow Unificationists had already devoted their lives to destroying Marxism," Wells confesses. "When Father chose me (along with about a dozen other seminary graduates) to enter a Ph.D. program in 1978, I welcomed the opportunity to prepare myself for battle." Wells went out and earned his doctorate and penned *The Icons of Evolution*.

Beyond the legal and religious angling, a motivation for Intelligent Design theorists to distance themselves from the creationists of old is that no one took the creationists seriously. Dembski, who has no qualms about contesting creationist beliefs of other stripes to further his cause, explained the problem in a 2005 debate with the Young Earth creationist Henry Morris. The first step, he says, involves "dismantling materialism. . . . Not only does intelligent design rid us of this ideology, which suffocates the human spirit, but, in my personal experience, I've found that it opens the path for people to come to Christ. Indeed, once materialism is no longer an option, Christianity again becomes an option." The objective then is to find a foothold for ridding the world of materialism. In Dembski's view, "intelligent design should be viewed as a ground-clearing operation that gets rid of the intellectual rubbish that for generations has kept Christianity from receiving serious consideration."[8]

The new creationism may differ in the details from the old creationism, but their ultimate goals run parallel. The veneer of science in Intelligent Design theory is there purposely to cover up the religious agenda. Indeed, when you press Intelligent Design creationists on what science, precisely, they are practicing, they admit in person that they have not yet developed "that part" of their program. For a similarly honest self-appraisal of the Intelligent Design movement recorded in print, Dembski's 2004 book, *The Design Revolution*, provides the money quote (or the confession): "Because of intelligent design's outstanding success at gaining a cultural hearing, the cultural and political component of intelligent design is now running ahead of the scientific and intellectual component."[9] At a 2004 meeting of the Bible Institute of Los Angeles, Paul Nelson confirmed Dembski's assessment. "Easily the biggest challenge facing the ID community is to develop a full-fledged theory of biological design," Nelson said. "We don't have such a theory right now, and that's a problem. . . . Right now, we've got a bag of powerful intuitions, and a handful of notions such as 'irreducible complexity' and 'specified complexity'—but, as yet, no general theory of biological design."[10]

The true measure of a scientific theory is whether any scientists use it or not, and no scientists are using Intelligent Design theory. Even vocally Christian scientists do not use the intuitions of Intelligent Design in place of the scientific method. Lee Anne Chaney, a professor of biology at the Christian-based Whitworth College, sums it up:

> As a Christian, part of my belief system is that God is ultimately responsible. But as a biologist, I need to look at the evidence. Scientifically speaking, I don't think intelligent design is very helpful because it does not provide things that are refutable—there is no way in the world you can show it's not true. Drawing inferences about the deity does not seem to me to be the function of science because it's very subjective.[11]

The Intelligent Design movement "does not provide things that are refutable" because its real objective is not to prove a scientific theory but to gain ground for religious ideology.

Follow the Money

Science or no science, to illuminate the agenda behind Intelligent Design we can employ the tried-and-true method of political analysis: Follow the money. According to an extensive investigation by *The New York Times*, the Seattle-based Discovery Institute—the nonprofit organization that has been the hammer of the Wedge movement—has been funded primarily by right-wing religious groups. The Ahmanson Foundation, for example, donated $750,000 through its executor, Howard Ahmanson, Jr., who once said his goal is "the total integration of biblical law into our lives." The MacLellan Foundation, a group that commits itself to "the infallibility of the Scripture" and gives grants to organizations "committed to furthering the Kingdom of Christ," donated $450,000 to the Discovery Institute. In 1998, Howard F. Ahmanson's conservative philanthropy, Fieldstead & Company, granted the Discovery Institute $300,000 per year for five years, and in 1999 the Stewardship Foundation increased its grant to $200,000 per year for five years. According to its Web site, the Stewardship Foundation was established "to contribute to the propagation of the Christian Gospel by evangelical and missionary work." Most of the other twenty-two foundations supporting the Discovery Institute with financial contributions were identified by the *Times* as politically conservative, including the Henry P. and Susan C. Crowell Trust of Colorado Springs, whose Web site describes its mission as "the teaching and active extension of the doctrines of evangelical Christianity," and the AMDG Foundation in Virginia, whose initials stand for *Ad Majorem Dei Gloriam*, Latin for "To the greater glory of God."

The *Times* also investigated the tax documents for the Discovery Institute and found that annual giving from conservative groups had increased from $1.4 million in 1997 to $4.1 million in 2003. With an annual budget of $3.6 million a year since 1996, the Discovery Institute has been sponsoring fellowships of $5,000 to $60,000 per year to fifty researchers. According to Stephen Meyer, a recipient of Discovery largess, 39 percent of the Discovery Institute's Center for Science and Culture budget of $9.3 million since 1996 has gone to underwrite various publishing projects.[12]

Money talks. At the time of this writing there are no fewer than seventy-eight pending legal clashes between Intelligent Design and evolution in thirty-one different states. Most of these have been fueled by the Discovery Institute's funding program. Fifty books, countless opinion editorials, essays, reviews, and commentaries, even slick documentaries—two broadcast on public television and one shown at the Smithsonian Institution—have also come down the pipeline.[13] As a poignant example of what money can buy, at the urging of the Discovery Institute's public relations firm—the same firm that promoted conservative congressman Newt Gingrich's 1994 Contract with America—in July 2005, Catholic cardinal Christof Schönborn wrote an opinion editorial in *The New York Times* in which he contradicted Pope John Paul II's 1996 statement that the theory of evolution is no threat to religion. Schönborn told Catholics that the Church does not accept evolution, a stunning reversal countered by the Vatican itself when Cardinal Paul Poupard held a press conference to declare that Genesis and evolution are "perfectly compatible."[14]

The Discovery Institute is about politics, not science. According to its president, Bruce Chapman, described by the *Times* as "a Rockefeller Republican turned Reagan conservative" who draws a hefty salary of $141,000 a year, "we are not going through this exercise just for the fun of it. We think some of these ideas are destined to change the intellectual—and in time the political—world." He is careful to add that "Fieldstead & Company and the Stewardship Foundation agree, or they would not have given us such substantial funding."[15] The Discovery Institute has become so political, in fact, that the Templeton Foundation—the provider of the largest cash prize available (over $1 million) for "progress in religion"—has withdrawn its support. After giving the Discovery Institute $75,000 for a 1999 conference on Intelligent Design, they have since rejected the institute's grant proposals. Why? "They're political—that for us is problematic," explained the senior vice president of the Templeton Foundation, Charles L. Harper, Jr., who added that although Discovery has "always claimed to be focused on the science, what I see is much more focused on public policy, on public persuasion, on educational advocacy and so forth."[16]

The Greater Glory

Although the motives of the proponents of Intelligent Design are secondary to their arguments, these motives are misplaced.

Let us reconsider the motto of the Christian AMDG Foundation—*Ad Majorem Dei Gloriam*—"To the greater glory of God." These are stirring words, even emblazoned on the stationery of Pope John Paul II, the same Pope who granted one billion Catholics permission to accept evolution as a reality of nature that poses no threat to religion.

If you are a theist, what could possibly proclaim the greater glory of God's creation more than science, the instrument that has illuminated more than any other tool of human knowledge the grandeur in this evolutionary view of life? There are questions that remain to be answered, to be sure, and controversies still to be resolved, but they are questions and controversies open to all of us—theists and atheists, conservatives and liberals—because science knows no religious or political boundaries. Science, more than any other tradition, follows the motto erected at the Panama Canal: *Aperire Terram Gentibus,* "To Open the World to All People."

7

WHY SCIENCE CANNOT
CONTRADICT RELIGION

———————————— ◆ ————————————

The sciences of observation describe and measure the multiple manifestations of life with increasing precision and correlate them with the time line. The moment of transition to the spiritual cannot be the object of this kind of observation.

—Pope John Paul II, *Truth Cannot Contradict Truth*, 1996

In considering the religious implications of the theory of evolution, it is illuminating to consider in greater depth the religious attitudes of the theory's architect. Charles Darwin's thoughts and feelings on how science and religion might be reconciled—in his own home and in the larger society—were complex and evolved over time.

Darwin matriculated at Cambridge University in theology, but he did so only after abandoning his medical studies at Edinburgh University because of his distaste for the barbarity of surgery. Darwin's famous grandfather Erasmus, and his father Robert, both physicians by trade who were deeply schooled in natural history, were confirmed freethinkers, so there was no doctrinaire pressure on the young Charles to choose theology.

In point of fact, Darwin's selection of theology as his primary course of study allowed him to pursue his passion for natural history through the academic justification of studying "natural theology"—

he was far more interested in God's works (nature) than God's words (the Bible). Besides, theology was one of only a handful of professions that a gentleman of the Darwin family's high social position in the landed gentry of British society could choose. Finally, although Darwin belonged to the Church of England, membership was expected of someone in his social class.

Still, Darwin's religiosity was not entirely utilitarian. He began and ended his five-year voyage around the world as a creationist, and he regularly attended religious services on board the *Beagle* and even during some land excursions in South America. It was only upon his return home that the loss of his faith came about, and that loss happened gradually—even reluctantly—over many years.

Darwin's God and the Devil's Chaplain

Nagging doubts about the nature and existence of the deity chipped away at Darwin's faith as a result of his studies of the natural world, particularly many of his observations of the cruel nature of the relationship between predators and prey. "What a book a Devil's Chaplain might write on the clumsy, wasteful, blundering low & horridly cruel works of nature!" Darwin lamented in an 1856 letter to his botanist mentor Joseph Hooker. In 1860 he wrote to his American colleague, the Harvard biologist Asa Gray, about a species of wasp that paralyzes its prey (but does not kill it), then lays its eggs inside the paralyzed insect so that upon birth its offspring can feed on live flesh. "I cannot persuade myself that a beneficent God would have designedly created the *Ichneumonidae* with the express intention of their feeding within the living bodies of Caterpillars, or that a cat should play with mice. Not believing this," he reflected, "I see no necessity in the belief that the eye was expressly designed."[1]

Pain and evil in the human world made Darwin doubt even more. "That there is much suffering in the world no one disputes," he wrote to a correspondent. "Some have attempted to explain this with reference to man by imagining that it serves for his moral improvement. But the number of men in the world is as nothing compared with that of all other sentient beings, and they often suffer greatly without any moral improvement." Which is more likely, that pain and evil are the result of an all-powerful and good God, or the product of uncaring natural forces? "The presence of much suffering agrees well with the view that all organic beings have been developed through variation and natural selection."[2] The death of Darwin's beloved ten-year-old daughter Anne put an end to whatever confidence he had in God's benevolence, omniscience, and even existence. According to the great Darwin scholar and biographer Janet Browne, "this death was the formal beginning of Darwin's conscious dissociation from believing in the traditional figure of God."[3]

Throughout most of his professional career, however, Darwin eschewed the God question entirely, choosing instead to focus on his scientific studies. Toward the end of his life Darwin received many letters querying him on his religious attitudes. His long silence gave way to a few revelations. In one letter dated 1879, just three years before he died, Darwin finally expressed his beliefs: "In my most extreme fluctuations I have never been an Atheist in the sense of denying the existence of God. I think that generally (and more and more as I grow older), but not always, that an Agnostic would be the more correct description of my state of mind."[4]

A year later, Darwin clarified his thinking. The British socialist Edward Aveling had compiled a volume entitled *The Student's Darwin*, on the implications of evolutionary theory for religious thought, and Aveling wanted Darwin's endorsement. The book had a militant antireligious flavor and unabashedly radical atheist tone that Darwin disdained, and he declined the request, elaborating his reason with his usual flair for quotable maxims: "It appears to me (whether rightly or wrongly) that direct arguments against christianity & theism produce hardly any effect on the public; & freedom of thought is best promoted by the gradual illumination of men's minds which follow[s] from the advance of science. It has, therefore, been always my object to avoid writing on religion, & I have confined myself to science." He then appended an additional hint about a personal motive, noting "I may, however, have been unduly biased by the pain which it would give some members of my family, if I aided in any way direct attacks on religion."[5] Darwin's wife Emma was a deeply religious woman, and out of respect for her he kept the public side of his religious skepticism in check, an admirable feat of self-discipline by a man of high moral character.

Conflict or Compromise?

Was Darwin's approach to science and religion healthy? Was it logical? Is it possible to reconcile religious belief with scientific thinking? The answer one gives to these questions determines the attitude one takes to the relationship of science and religion: conflict, harmony, or indifference. And if we could find some level at which agreement could be reached between all sides of the debate, much of the angst and rancor in today's culture over this divide would subside. I have made such an attempt in the form of a three-tiered model of the possible relationships between science and religion.

1. *The Conflicting-Worlds Model*. This "warfare" approach holds that science and religion are mutually exclusive ways of knowing,

one being right and the other wrong. In this view, the findings of modern science are always a potential threat to one's faith and thus they must be carefully vetted against religious truths before acceptance; likewise, the tenets of religion are always a potential threat to science and thus they must be viewed with skepticism and cynicism. The conflicting-worlds model is embraced by extremists on both sides of the divide. Young Earth creationists, who insist that all scientific findings must correlate perfectly with their own (often literal) reading of Genesis, retain a suspicious hostility toward science, while militant atheists cannot imagine how religion could contribute anything positive to human knowledge or social interaction.[6]

2. *The Same-World Model.* More conciliatory in its nature than the conflicting-worlds model, this position holds that science and religion are two ways of examining the same reality, and that as science progresses to a deeper understanding of the natural world, it will reveal that many ancient religious tenets are true. The same-world model is embraced by many mainstream theologians, religious leaders, and believing scientists who prefer a more flexible cognitive approach to science and religion, allowing them to read biblical passages metaphorically. For example, the "days" in the Genesis creation story may represent geological epochs of great length. The theology of Pope John Paul II as well as that of the Dalai Lama fall squarely into this tier, as they argue that science and religion can work together toward the same goal of understanding the universe and our place in it.[7]

3. *The Separate-Worlds Model.* On this tier, science and religion are neither in conflict nor in agreement but are, in Stephen Jay Gould's phrase, "nonoverlapping magisteria" (NOMA).[8] Before science began its ascent four centuries ago, religion provided an explanation for the natural world in the form of various cosmogony

myths. Since the scientific revolution, however, science has taken over the job of explaining the natural world, making obsolete ancient religious sagas of origins and creation. Yet religion thrives in the modern age because it still serves a useful purpose as an institution for social cohesiveness and as a guide to finding personal meaning and spirituality, a function that science has left largely untouched.

God as a Null Hypothesis

Can the conflicting-worlds and same-world models of science and religion work? Frankly, they cannot. To accept science requires accepting one of its central tenets: that a claim must be *falsifiable;* that is, there has to be some way to test the claim and show that it is false. If it cannot be proven false, then it cannot be proven true. The philosopher of science Karl Popper made the definitive statement on the matter: "I shall not require of a scientific system that it shall be capable of being singled out, once and for all, in a positive sense; but I shall require that its logical form shall be such that it can be singled out, by means of empirical tests, in a negative sense: it must be possible for an empirical scientific system to be refuted by experience."[9]

On the question of God's existence, what criteria for falsifiability could we establish? If we want to make God's existence a scientific question that can be decided by empirical evidence, we would need to establish an operational definition of God and quantifiable criteria by which we can arrive at a testable conclusion of the deity's existence. In experimental science we begin by accepting the "null hypothesis" that whatever is being tested does not exist or has no effect. If the evidence is significant, we may "reject the null hypothesis" and conclude that our subject does exist or has some effect. In subjecting God to experimental science, we would have to begin by accepting the null hypothesis that He does not exist, and then assess the evidence to determine if it is significant enough to reject the null hypothesis.

The claim that intercessory prayer (in which one prays for God to intercede) can effect healing, for example, is testable. If true, it would imply that the deity is acting in our world in some measurable fashion. However, the handful of studies that have found significant differences between the prayed-for experimental group and the not-prayed-for control group have had deep methodological flaws (such as not controlling for age, socioeconomic class, or condition of health before entering the hospital, all of which influence recovery).[10] To date, strictly controlled prayer studies, as a testable hypothesis of God's divine providence, have failed the test.

The numerous other claims by Intelligent Design creationists that science supports belief in God also fall dramatically short of the empirical standards of science. Based on these results, were we to take a strictly scientific approach to the God question, we would have to reject the God hypothesis. Are theists willing to go this far when they attempt to use science to support their religious tenets? I doubt it, which is precisely why the separate-worlds model is the best approach to take for theists.

A Is A: Why Science Cannot Contradict Religion

Darwin's separate-worlds approach to science and religion worked well for him in both his home and his culture, but it still leaves open the deeper question about whether one can logically believe in God and accept evolution. That is, if carried to its logical conclusion, does the theory of evolution preclude belief in God? This is where the epistemological rubber meets the hypothetical road.

Belief in God depends on religious faith. Acceptance of evolution depends on empirical evidence. This is the fundamental difference between religion and science. If you attempt to reconcile and combine religion and science on questions about nature and the universe, and if you push the science to its logical conclusion, you will end up naturalizing the deity; for any question about nature, if your answer is "God did it," a scientist will ask such questions as "*How* did God do it? What *forces* did God use? What forms of *matter* and *energy* were employed in the creation process?" The end result of this inquiry can only be natural explanations for all natural phenomena. What place, then, for God?

One could logically argue that God *is* the laws and forces of nature, but this is pantheism and not the type of personal God in which most people profess belief. One could also reasonably argue that God created the universe and life using the laws and forces of nature, but it leaves us with those nagging scientific questions: *Which* laws and forces were used, and in *what manner* were they used? For that matter, *how* did God create the laws and forces of nature? A scientist would be curious to know God's recipe for, say, gravity. Likewise, it is a legitimate scientific question to ask: *What* made God, and *how* was God created? How do you make an omniscient and omnipotent being?

The theists' response to this line of inquiry is that God needs no cause—God is a causeless cause, an unmoved mover. But why should God not need a cause? If the universe is everything that is, ever was, or ever shall be, God must be within the universe or *be* the universe. In either case, God would himself need to be caused, and thus the regress to a first cause leads back to the question: What caused God? And if God does not need to be caused, then clearly not everything in the universe needs to be caused. Maybe the initial creation of the universe was its own first cause and the Big Bang was the prime mover.

The problem with all of these attempts at blending science and religion may be found in a single principle: *A is A*. Or: *Reality is real*. To attempt to use nature to prove the supernatural is a violation of *A is A*. It is an attempt to make reality unreal. *A* cannot also be *non-A*. Nature cannot also be non-nature. Naturalism cannot also be supernaturalism.

Pope John Paul II, whose theology was influenced by Aristotle and Aquinas, two of the greatest minds in the history of philosophy and theology, understood this fundamental principle and argued the point in his 1996 encyclical, *Truth Cannot Contradict Truth*. The only way science and religion can be reconciled, particularly in the context of the evolution-creation controversy, is if body and soul are ontologically distinct; that is, if they exist in different realities. Evolution created the body, God created the soul:

> With man, then, we find ourselves in the presence of an ontological difference, an ontological leap, one could say. Consideration of the method used in the various branches of knowledge makes it possible to reconcile two points of view which would seem irreconcilable. The sciences of observation describe and measure the multiple manifestations of life with increasing precision and correlate them with the time line. The moment of transition to the spiritual cannot be the object of this kind of observation, which nevertheless can discover at the experimental level a series of very valuable signs indicating what is specific to the human being. But the experience of metaphysical knowledge, of self-awareness and self-reflection, of moral conscience, freedom, or again, of aesthetic and religious experience, falls within the competence of philosophical analysis and reflection, while theology brings out its ultimate meaning according to the Creator's plans.[11]

Believers can have both religion and science as long as there is no attempt to make *A* *non-A,* to make reality unreal, to turn naturalism into supernaturalism. Thus, the most logically coherent argument for theists is that *God is outside time and space;* that is, God is beyond nature—super nature, or supernatural—and therefore cannot be explained by natural causes. God is beyond the dominion of science, and science is outside the realm of God.

8

WHY CHRISTIANS AND CONSERVATIVES SHOULD ACCEPT EVOLUTION

———————————— ◆ ————————————

I see no good reason why the views given in this volume should shock the religious feeling of any one. It is satisfactory, as showing how transient such impressions are, to remember that the greatest discovery ever made by man, namely, the law of the attraction of gravity, was also attacked by Leibnitz, "as subversive of natural, and inferentially of revealed, religion." A celebrated author and divine has written to me that "he has gradually learnt to see that it is just as noble a conception of the Deity to believe that He created a few original forms, capable of self-development into other and needful forms, as to believe that He required a fresh act of creation to supply the voids caused by the actions of His laws."

—Charles Darwin, 2nd edition of *On the Origin of Species,* 1860

During the cultural brouhaha whipped up by the media frenzy over President George W. Bush's 2005 comments on Intelligent Design and evolution, a reporter from *Time* magazine solicited my opinion about whether one can believe in both God and evolution.

I replied that, empirically speaking, apparently so, because lots of people do—a 1996 survey found that 39 percent of American scientists profess belief in God, and a 1997 poll found that 99 percent of American scientists accept the theory of evolution. More recently, preliminary results from a long-term survey of 1,600 scientists from twenty-one elite universities revealed that over half consider themselves "moderately spiritual" to "very spiritual," and about a third hold formal religious affiliations.[1] So either a third of my colleagues live in a cognitive fantasyland of logic-tight compartments, or there is a way to find that separate-worlds harmony between science and religion.

If scientists can believe in God and evolution, can Christians? Using the same empirical standards, evidently so, because approximately 96 million American Christians do: In a 2001 Gallup poll, 37 percent of Americans (107 million people) agreed with the statement "Human beings have developed over millions of years from less advanced forms of life, but God guided this process." Since about 90 percent of Americans are Christians, approximately 96 million Christians believe that God used evolution to guide the process of creating advanced forms of life.[2]

Even many evangelical Christians—the religious cohort most outspoken against the theory—accept evolution. Consider the statement by former president Jimmy Carter—who identifies himself as an evangelical Christian—in response to a measure passed in Georgia in 2004 that required all public school biology textbooks to include a sticker proclaiming:

> This textbook contains material on evolution. Evolution is a theory, not a fact, regarding the origin of living things. This material should be approached with an open mind, studied carefully and critically considered.

President Carter was outraged. "As a Christian, a trained engineer and scientist, and a professor at Emory University, I am embarrassed by Superintendent Kathy Cox's attempt to censor and distort the education of Georgia's students," he wrote. "The existing and long-standing use of the word 'evolution' in our state's text-books has not adversely affected Georgians' belief in the omnipotence of God as creator of the universe. There can be no incompatibility between Christian faith and proven facts concerning geology, biology, and astronomy. There is no need to teach that stars can fall out of the sky and land on a flat Earth in order to defend our religious faith."[3] The requirement was subsequently repealed, though not before it served as fodder for other state legislatures as well as for political cartoonists.

And as seen in the previous chapter, the compatibility of God and Darwin finds evidence in the one billion Catholics who embraced Pope John Paul II's 1996 Pontifical Academy of Sciences Encyclical. He asserts that evolution happened, that it is okay to accept it as fact, and that it is no threat to religion:

> New knowledge has led to the recognition that the theory of evolution is more than a hypothesis. It is indeed remarkable that this theory has been progressively accepted by researchers, following a series of discoveries in various fields of knowledge. The convergence, neither sought nor fabricated, of the results of work that was conducted independently is in itself a significant argument in favor of the theory.[4]

Christians, Conservatives, and Evolution

Despite these examples, recent polling data show that we still have a way to go before the theory of evolution achieves total acceptance. According to a 2005 Pew Research Center poll, 70 percent of evangelical Christians believe that living beings have always existed in their present form, compared to 32 percent of Protestants and 31 percent of Catholics; politically, 60 percent of Republicans are creationists while only 11 percent accept evolution, compared to 29 percent of Democrats who are creationists and 44 percent who accept evolution. Similarly, a 2005 Harris poll found that 63 percent of liberals but only 37 percent of conservatives believe that humans and apes have a common ancestry; further, those with a college education, those between the ages of eighteen and fifty-four, and those from the Northeast and West are more likely to accept evolution, whereas those without a college degree, aged fifty-five and older, and from the South are more likely to believe in creationism.[5]

What these figures tell us is that the nonscientific, demographic reasons for rejecting evolution, most notably religion and politics, are very strong. Can a Christian be a Darwinian? Can a conservative accept evolution? Yes. Here is how and why.

Evolution Makes for Good Theology

As outlined in this book, the theory describing how evolution happened is one of the most well-founded in all of science. Christians and conservatives embrace the value of truth-seeking as much as non-Christians and liberals do, so evolution should be accepted by everyone because it is true. In this sense, evolution is no different from any other scientific theory already fully accepted by both Christians and conservatives, such as heliocentrism, gravity, continental drift and plate tectonics, the germ theory of disease, the genetic basis of heredity, and many others.

Christians believe in a God who is omniscient, omnipotent, and eternal. The glory of the creation commands reverence regardless of when creation took place. And compared to omniscience and omnipotence, what difference does it make *how* God created life—via spoken word or via natural forces? The grandeur of life's complexity elicits awe regardless of what creative processes were employed. Christians should embrace modern science for what it has done to reveal the magnificence of the divinity in a depth and detail unmatched by ancient texts.

In contrast, Intelligent Design creationism reduces God to an artificer, a mere watchmaker piecing together life out of available parts in a cosmic warehouse. If God is a being in space and time, it means that He is restrained by the laws of nature and the contingencies of chance, just like all other beings of this world. An omniscient and omnipotent God must be above such constraints, not subject to nature and chance. God as creator of heaven and earth and all things visible and invisible would need necessarily to be outside such created objects. If He is not, then God is like us, only smarter and more powerful; but not omniscient and omnipotent. Calling God a watchmaker is delimiting.

But more important, evolution explains family values and social harmony. Humans and other social mammals, including and especially apes, monkeys, dolphins, and whales, share a host of characteristics: attachment and bonding, cooperation and mutual aid, sympathy and empathy, direct and indirect reciprocity, altruism and reciprocal altruism, conflict resolution and peace-making, community concern and reputation caring, and awareness of and response to the social rules of the group. As a social primate species we evolved the capacity for positive moral values because they enhance the survival of both family and community. Evolution created these values in us, and religion identified them as important in order to accentuate them. "The following proposition seems to me in a high degree probable," Darwin theorized in *The Descent of Man,* "namely, that any animal whatever, endowed with well marked social instincts, the parental and filial affections being here included, would inevitably acquire a moral sense or conscience, as soon as its intellectual powers had become as well, or nearly as well developed, as in man." The evolution of the moral sense was a step-wise process, "a highly complex sentiment, having its first origin in the social instinct, largely guided by the approbation of our fellow-men, ruled by reason, self-interest, and in later times by deep religious feelings, confirmed by instruction and habit, all combined, constitute our moral sense and conscience."[6]

Evolution also explains evil, original sin, and the Christian model of human nature. We may have evolved to be moral angels, but we are also immoral beasts. Whether you call it evil or original sin, humans have a dark side. Individuals in our evolutionary ancestral environment needed to be both cooperative and competitive, for example, depending on the context. Cooperation leads to more successful hunts, food sharing, and group protection from predators and enemies. Competition leads to more resources for oneself and family, and protection from other competitive individuals who are less inclined to cooperate, especially those from other groups. Thus we are by nature both cooperative and competitive, altruistic and selfish, greedy and generous, peaceful and bellicose; in short, good and evil. Moral codes, and a society based on the rule of law, are necessary not just to accentuate the positive, but especially to attenuate the negative side of our evolved nature. Christians would find little to disagree with in the observation of Thomas Henry Huxley, Darwin's chief defender in the nineteenth century: "Let us understand, once for all, that the ethical process of society depends, not on imitating the cosmic process, still less in running away from it, but in combating it."[7]

Thus, by explaining the origins of our positive and negative behaviors and characteristics, evolution explains the origin of morality and religions designed moral codes based on our evolved natures. For the first ninety thousand years of our existence as fully modern humans, our ancestors lived in small bands of tens to hundreds of individuals. In the last ten thousand years, these bands evolved into tribes of thousands; tribes developed into chiefdoms of tens of thousands; chiefdoms coalesced into states of hundreds of thousands; and states conjoined into empires of millions. How and why did this happen? By ten thousand years ago, our species had spread to nearly every region of the globe and people everywhere lived where they could hunt and gather. This system tended to contain populations, but the invention of agriculture around that time allowed these populations to explode. With those increased populations came new social technologies for governance and conflict resolution: politics—and religion.

The moral emotions, such as guilt and shame, pride and altruism, evolved in those tiny bands of one hundred to two hundred people as a form of social control and group cohesion. One means of accomplishing this was through reciprocal altruism—"I'll scratch your back if you'll scratch mine." But as Lincoln noted, men are not angels. People defect from informal agreements and social contracts. In the long run, reciprocal altruism works only when you know who will cooperate and who will defect. This information is gathered in various ways, including through stories about other people—more commonly known as gossip. Most gossip is about relatives, close friends, those in our immediate sphere of influence, and members of the community or society who have high social status. It is here we find our favorite subjects of gossip: sex, generosity, cheating, aggression, social status and standings, births and deaths, political and religious commitments, and the various nuances of human relations, particularly friendships and alliances.

When bands and tribes gave way to chiefdoms and states, religion developed as a principal social institution to accentuate amity and attenuate enmity. It did so by encouraging altruism and selflessness, discouraging excessive greed and selfishness, and especially by revealing the level of commitment to the group through social events and religious rituals. If I see you every week participating in our religion's activities and following the prescribed rituals, this is an indication that you can be trusted. As organizations with codified moral rules and the power to enforce the rules and punish their transgressors, religion and government responded to a need.

Consider the biblical command to "love thy neighbor." In the Paleolithic social environment in which our moral sentiments evolved, one's neighbors were family, extended family, and community members who were well known to everyone. To help others was to help oneself. In chiefdoms, states, and empires, the decree meant one's immediate in-group. Out-groups were not included. This explains the seemingly paradoxical nature of Old Testament morality, where on one page high moral principles of peace, justice, and respect for people and property are promulgated, and on the next page raping, killing, and pillaging people who are not one's "neighbors" are endorsed. Deuteronomy 5:17, for example, admonishes, "Thou shalt not kill," yet in Deuteronomy 20:10–18, the Israelites are commanded to lay siege to an enemy city, steal the cattle, enslave those men who surrender, and kill those who do not.

The cultural expression of this in-group morality is not restricted to any one religion, nation, or people. It is a universal human trait common throughout history, from the earliest bands and tribes to modern nations and empires. Christian morality, like that of many other religions, was designed to help us overcome these natural tendencies.

Much of Christian morality has to do with human relationships, most notably truth telling and sexual fidelity, because the violation of these causes a severe breakdown in trust, and once trust is gone there is no foundation on which to build a family or a community. Evolution explains why. We evolved as pair-bonded primates for whom monogamy is the norm (or, at least, serial monogamy—a sequence of monogamous relationships). Adultery is a violation of a monogamous bond, and there are copious scientific data showing how destructive adulterous behavior is to a monogamous relationship. (In fact, one of the reasons that "serial monogamy" best describes the mating behavior of our species is that adultery typically destroys a relationship, forcing couples to split up and start over with someone new.) This is why most religions are unequivocal on the subject. Consider Deuteronomy 22:22: "If a man is found lying with the wife of another man, both of them shall die, the man who lay with the woman, and the woman; so you shall purge the evil from Israel."

Most religions decree adultery to be immoral, but this is because evolution made it immoral. According to evolutionary psychologist David Buss, sexual betrayals are primarily a biologically driven phenomenon encoded over eons of Paleolithic cuckolding. Buss argues that there are differences between men and women in this tendency, and that these differences hold across different cultures; thus, they are primarily driven by our genes. In one study by psychologists Russell Clark and Elaine Hatfield, an attractive member of the opposite sex posed one of three questions to fellow single college students:

1. "Would you go out on a date with me tonight?"
2. "Would you go back to my apartment with me tonight?"
3. "Would you sleep with me tonight?"

The results were revealing. For women, 50 percent agreed to the date, 6 percent agreed to return to the apartment, and not a single one of them agreed to have sex. By contrast, for men, 50 percent agreed to the date, 69 percent agreed to the apartment, and 75 percent agreed to the sex! No wonder most religions have strict codes of sexual restraint against men and repeated warnings to women about the power of the sex drive.[8]

As for the act of adultery itself, its evolutionary benefits are obvious. For the male, depositing his genes in more places increases the probability of his genes making it into the next generation. For the female, it is a chance to trade up for better genes, greater resources, and higher social status. The evolutionary hazards of adultery, however, often outweigh the benefits. For the male, revenge by the adulterous woman's husband can be extremely dangerous, if not fatal—a significant percentage of homicides involve love triangles. And while getting caught by one's own wife is not likely to result in death, it can result in loss of contact with children, loss of family and security, and risk of sexual retaliation, thus decreasing the odds of one's mate bearing one's own offspring. For the female, being discovered by the adulterous man's wife involves little physical risk, but getting caught by one's own husband can and often does lead to extreme physical abuse and occasionally even death. So evolutionary theory explains the origins and rationale behind the religious precept against adultery.

Likewise for truth telling and lying. Truth telling is vital for building trust in human relations, so lying is a sin. Unfortunately, research shows that all of us lie every day, but most of these are so-called "little white lies," in which we might exaggerate our accomplishments, or lies of omission, in which information is omitted to spare someone's feelings or save someone's life—if an abusive husband inquires whether you are harboring his terrified wife, it would be immoral for you to answer truthfully. Such lies are usually considered amoral. Big lies, however, lead to the breakdown of trust in personal and social relationships, and these are considered immoral. Evolution created a system of deception detection because of the importance of trusting social relations to our survival and fecundity. Although we are not perfect lie detectors (and thus you can fool some of the people some of the time), if you spend enough time and have enough interactions with someone, their honesty or dishonesty will be revealed, either through direct observation or by indirect gossip from other observers.[9] Thus, it is not enough to fake doing the right thing in order to fool our fellow group members, because although we are good liars, we are also good lie detectors. The best way to convince others that you are a moral person is not to fake being a moral person but actually to *be* a moral person. Don't just pretend to do the right thing, *do* the right thing. Such moral sentiments evolved in our Paleolithic ancestors living in small communities. Subsequently, religion identified these sentiments, labeled them, and codified rules about them.

Evolution and the Conservative Theory of Free Market Economics

Political conservatives can also find explanations—and foundations—in the theory of evolution. Charles Darwin's theory of *natural selection* is precisely parallel to Adam Smith's theory of the *invisible hand*. Darwin focused on showing how complex design and ecological balance were unintended consequences of individual competition among organisms. Smith focused on showing how national wealth and social harmony were unintended consequences of individual competition among people. The natural economy mirrors the artificial economy. Conservatives embrace free market capitalism, and they are against excessive top-down governmental regulation of the economy; they understand that the most efficient economy emerges from the complex, bottom-up behaviors of individuals pursuing their own self-interest without awareness of the larger consequences of their actions.

Adam Smith was a professor of moral philosophy who posited a theory of human nature with competing motives: We are both competitive and cooperative, altruistic and selfish. There are times of need when we can count on the humanity of strangers to help us, but daily trade in a marketplace is founded on the lesser angels of our natures. As Smith explained in *The Wealth of Nations,* "It is not from the benevolence of the butcher, the brewer, or the baker that we expect our dinner, but from their regard to their own interest. We address ourselves, not to their humanity but to their self-love, and never talk to them of our own necessities but of their advantages."[10] By allowing individuals to follow their natural inclination to pursue their self-love, the country as a whole will prosper, almost as if the entire system were being directed by . . . yes . . . an invisible hand. It is here that we find the one and only use of the metaphor in *The Wealth of Nations:*

> Every individual is continually exerting himself to find out the most advantageous employment for whatever capital he can command. . . . He generally, indeed, neither intends to promote the public interest, nor knows how much he is promoting it. He intends only his own security; and by directing that industry in such a manner as its produce may be of the greatest value, he intends only his own gain, and he is in this, as in many other cases, led by an *invisible hand* to promote an end which was no part of his intention. By pursuing his own interest he frequently promotes that of the society more effectually than when he really intends to promote it.[11]

Compare this to Darwin's description of what happens in nature when organisms pursue their self-love with no cognizance of the unintended consequences of their behavior:

> It may be said that *natural selection* is daily and hourly scrutinising, throughout the world, every variation, even the slightest; rejecting that which is bad, preserving and adding up all that is good; silently and insensibly working, whenever and wherever opportunity offers, at the improvement of each organic being in relation to its organic and inorganic conditions of life. We see nothing of these slow changes in progress, until the hand of time has marked the long lapses of ages, and then so imperfect is our view into long past geological ages, that we only see that the forms of life are now different from what they formerly were.[12]

Inheriting the Wind

Evolution provides a scientific foundation for the core values shared by most Christians and conservatives, and by accepting—and embracing—the theory of evolution, Christians and conservatives strengthen their religion, their politics, and science itself. The conflict between science and religion is senseless. It is based on fears and misunderstandings rather than on facts and moral wisdom. Indeed, for the benefit of our society, the battle currently being played out in curriculum committees and public courtrooms over evolution and creationism must end now, or else, as the book of Proverbs (11:29) warned:

> He that troubleth his own house shall inherit the wind: and the fool shall be servant to the wise of heart.

9

THE REAL UNSOLVED
PROBLEMS IN EVOLUTION

——————————————◆——————————————

There are known knowns. There are things we know we know.
We also know there are known unknowns. That is to say, we
know there are some things we do not know. But there are also
unknown unknowns, the ones we don't know we don't know.

—Donald Rumsfeld, United States Secretary of Defense,
press conference statement on February 12
(Darwin's birthday), 2002

On September 18, 1835, H.M.S. *Beagle* dropped anchor in the
Galápagos archipelago at the base of Frigatebird Hill on Chatham
Island, now known as San Cristóbal. Blue-footed boobies circled
about the bay, and at the appropriate moment tucked their wings
back and sliced into the shallow sea to scoop their prey from
schools of thousands of tiny fish. Earning their moniker, the
frigatebirds perched high on the cliffs above, poised like pirates to
pounce on the boobies and steal their catch.

Charles Darwin's first impression of this island was "what we
might imagine the cultivated parts of the Infernal regions to be."
Vast swatches of black volcanic rock and countless extinct cinder
cones were punctuated with scrappy life forms suited for here and

nowhere else. Most striking to Darwin were the marine iguanas that swarmed the rocky beaches of the northern regions of the island:

> The black Lava rocks on the beach are frequented by large (2–3 ft) most disgusting, clumsy Lizards. They are as black as the porous rocks over which they crawl & seek their prey from the sea.—Somebody calls them "imps of darkness."—They assuredly well become the land they inhabit.[1]

When Frank Sulloway and I hit the beaches of San Cristóbal 170 years later, we searched in vain for Darwin's imps. In their stead we spotted feral cats darting in and out of the black boulders, the largest, fastest, and stealthiest cats imaginable. The adult marine iguanas are too large and leathery for the fugitive felines, but the juveniles make easy targets. Without a juvenile cohort to maintain a viable breeding population, the iguanas suffered geographic extinction.

Frank and I reported this glum news to our colleagues at the 2005 World Summit on Evolution that was being held on a coastal outskirt of the lively little fishing town of Puerto Baquerizo Moreno, adjacent to Frigatebird Hill.[2] With 210 of the world's leading evolutionary biologists in attendance, the conference illuminated the greatest unsolved mysteries of evolution.

The Known and the Unknown

In the 1960s, Secretary of Defense Robert McNamara besieged the American public with eye-blurring statistical charts and graphs to demonstrate that we were winning a war in Vietnam that we were actually losing. Four decades later, Secretary of Defense Donald

Rumsfeld attempted the same sleight of hand, when he employed his infamous epistemology of known knowns, known unknowns, and unknown unknowns to explain the apparent nonexistence of Iraqi weapons of mass destruction.[3]

Creationists, Intelligent Design theorists, and outsiders to science often mistake the latter two categories for signs that the theory of evolution is in trouble, or that contentious debate between what we know and do not know means that the theory is false. Evolution is rich in controversy and disputation over the known and unknown. In reviewing what is on the cutting edge of debate within evolutionary biology, we discover the real questions we should be asking about evolution. If Intelligent Design creationists want to "teach the controversy," here are just a few of the major questions scientists are asking—and hoping to answer—about the origin and evolution of life.

How Did Life Begin and What Is the Origin of DNA?

Creationists revel in "how did it all begin?" questions, and the opening session of the Evolution Summit was on the origins of life, starting with a lecture by Antonio Lazcano, President of the International Society for the Study of the Origin of Life and a scientist at the Universidad Autónoma de México. Lazcano theorized that there were three sources for the primordial soup: volcanic outgassing, high-temperature submarine vents and fumaroles; and outer space—the 4.6-billion-year-old Murchison meteorite, discovered in Australia in 1969, for example, was loaded with such chemical building blocks of life as amino acids, aliphatic and aromatic hydrocarbons, hydroxy acids, purines, and pyrimidines. "The evidence strongly suggests that prior to the origin of life the primitive Earth already had many different catalytic agents, polymers with sequences of nucleotides, and membrane-forming compounds," Lazcano inferred, concluding that this prebiotic soup led to the first replicators, most likely RNA, and this led to the more complicated DNA replicator of today.

In his commentary on Lazcano's lecture, the UCLA paleobiologist William Schopf, *pace* Rumsfeld, asked: "What do we know? What are the unsolved problems? What have we failed to consider?" Schopf answered himself: "We know the overall sequence of life's origin, from CHONSP [carbon, hydrogen, oxygen, nitrogen, sulfur, phosphorus], to monomers, to polymers, to cells; we know that the origin of life was early, microbial, and unicellular; and we know that an RNA world preceded today's DNA-protein world. We do not know the precise environments of the early earth in which these events occurred; we do not know the exact chemistry of some of the important chemical reactions that led to life; and we do not have any knowledge of life in a pre-RNA world." As for what we have failed to consider, Schopf suggested that the "'pull of the present' makes it extremely difficult for us to model the early earth's atmosphere and the biochemistry of early life."

Later, Lynn Margulis, in her inimitable rapid-fire style, hit Lazcano with a point-blank question: "In your opinion, what came first, cells or the RNA world?" Lazcano answered: "If you define a cell as a membrane-enclosed system, then lipids-enclosed systems assisted in the polymerization of molecules, which led to RNA." Cells first, replicators second.

What Caused the Cambrian
"Explosion" of Life?

One of the creationists' favorite tactics is to focus on gaps in the fossil record, and there is no bigger gap, they claim, than the so-called "Cambrian explosion," a geologic period starting around 540 million years ago during which many of life's major taxa first made their appearance in the fossil record. Intelligent Design theorist Stephen Meyer portrays this "explosion" as a single and instantaneous event, rather than the reality of its development over a fifteen- to twenty-million-year period. (Paleontologists usually modify their description of the Cambrian "explosion" by noting that this is a "geological" moment that, by comparison to biological or human time, is glacially slow.) In fact, Meyer's paper challenging the standard evolutionary model of the Cambrian explosion was the first creationist paper ever published in a peer-reviewed scientific journal, and creationists have made much of that fact.[4]

Too bad Meyer was not present when Mikhail Fedonkin, head of the Laboratory of Precambrian Organisms at the Paleontological Institute in Moscow, reviewed the evidence for evolution in the billions of years preceding the Cambrian period. Fedonkin suggests that a fall of global temperatures, and the oxygenation of the biosphere caused by photosynthesis, played major roles in the dramatic change in the availability of heavy metals that he believes were crucial in the metabolic processes that led to the evolution of complex life. This metal-rich environment served as a catalyst: "Over 70 percent of known enzymes contain metal ions as a cofactor of an active site. Fast catalyzed reactions segregated life first dynamically and then structurally from the mineral realm." Once simple prokaryote cells gave rise to complex eukaryote cells, life was off and running, resulting quite naturally in the Cambrian eruption of complex hard-bodied organisms.

Commenting on Fedonkin's findings, Stefan Bengtson from the Swedish Museum of Natural History asked, "Why did the build-up to the Cambrian 'explosion' take so long?" Noting that 99.99999 percent of all species that ever lived have gone extinct, Bengtson answered his own question: "We do not know because we have nothing else to go on. Life is an evolutionary bush, not an evolutionary tree, but our data based on extant life induce us to prune the bush into a tree, so we need more data."

What Causes Major Shifts in Evolution?

Richard Fortey from the British Museum of Natural History investigates the importance of mass extinction events in resetting the direction of evolution, the magnitude of evolutionary arms races in driving morphological innovation, the relationship of climate change to evolutionary change, and the extent to which evolution can be described as directional. With half a billion years of a rich fossil record, Fortey said, we can track evolutionary periods of creativity and crises. Stephen Jay Gould's 1989 book *Wonderful Life* stimulated a lot of new ideas about the Cambrian explosion of life's diversity, he continued, and it soon became clear that there were an extensive variety of organisms difficult to classify, such as those found in the Cambrian-era Burgess Shale. But there are a number of Cambrian fossil beds, such as in China, where important phyla such as Chordata evolved. "In the Cambrian, some claim that there were as many as a hundred phyla, but the evidence does not support this. We now believe that morphological diversity did not explode as much as Gould originally suggested, although the explosion in evolutionary experimentation was real. By the time we get to the Cambrian, as at the Burgess Shale, the systems are very complex, such as trilobite eyes. Evolution was experimenting with many wondrous varieties, such as all the armor on the heads of trilobites."

What Is the Origin of Complex Life?

Intelligent Design theorists argue that evolutionary theory cannot account for the increase in organismal complexity. That is, how can we explain the increase of information in something like the genome in a world filled with entropy and the decay of information? This problem is addressed by Peter Gogarten, a professor of molecular and cell biology at the University of Connecticut, who demonstrates how often prokaryote organisms (simple cells like bacteria) experience horizontal gene transfer between organisms. They swap genes! "Over long periods of time gene transfer makes organisms existing in the same environment more similar to one another. This is most clearly seen in the case of organisms that live in environments that are otherwise inhabited by distant relatives only." Thus, Gogarten concluded, "the boundaries between prokaryotic species are fuzzy. Therefore the principles of population genetics need to be broadened so that they can be applied to higher taxonomic categories."

If organisms can swap genes, they can also acquire genes and thereby increase the information complexity of their genomes. If Lynn Margulis is right, this is in fact how simple prokaryote cells evolved into complex eukaryote cells (of which we are made), through symbiogenesis, which she described succinctly as "the inheritance of acquired genomes" and more formally in its relationship to symbiosis as "the long-term physical association between members of different types (species)." The problem with making evolution primarily about genetics (neo-Darwinism), Margulis concluded, is that "[r]andom changes in DNA alone do not lead to speciation. Symbiogenesis—the appearance of new behaviors, tissues, organs, organ systems, physiologies, or species as a result of symbiont interaction—is the major source of evolutionary novelty in eukaryotes: animals, plants, and fungi."

Creationists and Intelligent Design theorists also like to claim that the theory of evolution is a faith-based religion to which scientists must swear allegiance in order to obtain research grant money. They should have heard Margulis bellowing from the dais that neo-Darwinism is dead. Echoing Darwin, she said, "It was like confessing a murder when I discovered I was not a neo-Darwinist." But, she quickly added, "I am definitely a Darwinist. I think we are missing important information about the origins of variation. I differ from the neo-Darwinian bullies on this point." Neo-Darwinists tend to focus on plants and animals, Margulis complained. "We live on a bacterial planet," and evolutionary theory must be able to account for the evolution of the cell, "the fundamental unit of life," and all that a cell contains: "A minimal cell has DNA, mRNA, tRNA, rRNA, amino acylating enzymes, polymereses, sources of energy and electrons, lipoprotein membranes, and ion channels, all contained within a cell wall, and is an autopoietic [self-regulating feedback] system." The theory of symbiogenesis does just that, although there are scientists who remain skeptical.

How Many Branches Are There on the Human Evolutionary Tree?

Creationists' demand for just one transitional fossil is most notably made when it comes to human evolution, for which they claim there are none. They should have been at the lecture by the University of California, Berkeley, paleoanthropologist Timothy White, which he opened with a prediction made by Stephen Jay Gould in the late 1980s: "We know about three coexisting branches of the human bush. I will be surprised if twice as many more are not discovered before the end of the century." A glance at the extant fossil record today suggests that Gould was right. There are at least two dozen fossil species in six million years of hominid evolution.

But the bush is not so bushy, says White. The problem lies in the difference between "lumpers" and "splitters" in species classification, and in the social pressures to publish extraordinary new discoveries. If you want to get your fossil find published in *Science* or *Nature*, and you want the cover illustration, you cannot conclude that your fossil is yet another *Australopithecus africanus*, for example. You had better come up with an interpretation indicating that this new find you are revealing to the world for the first time is the most spectacular discovery of the last century and that it promises to overturn hominid phylogeny and send everyone back to the drawing board to reconfigure the human evolutionary tree. Training a more skeptical eye on these fossils, however, shows that many of them belong in already well established categories. White says that the specimen labeled *Kenyanthropus platyops*, for example, is very fragmented and is most likely just another *Australopithecus africanus*. "Name diversity does not equal biological diversity," White elucidated.

White concluded his talk with a fascinating discussion of the recent discovery of fossil dwarf humans on Flores Island in the Malay Archipelago, located on the outside of Wallace's Line, meaning that even during the last ice age they could have gotten there only by boat. (White did note, however, that after the 2004 tsunami in this area, people were rescued from large floating rafts of natural debris, so it is possible that the founding population of Flores humans rafted there by accident and not by design.) Found in Liang Bua cave, the type specimen of *Homo floresensis* was dated at 18,000 years old, meaning that they had to be modern humans because all other hominid species had long ago gone extinct. But with a cranial capacity of only 300cc—about the same size as that of Lucy or a modern chimpanzee—they were able to fashion complex tools (and possibly boats) with tiny brains; the implication is that brain architecture, not size, is what counts for higher intelligence. A second published specimen put to rest the pathology hypothesis that *Homo floresensis* was a microcephalic human. The best evidence, says White, points to insular dwarfing, a rapid punctuation event that led to the shrinkage of these isolated people. Such dwarfing effects can be seen on this and other islands, where large mammals get smaller (like the dwarf elephant) and small reptiles get larger (like the Komodo Dragon). The chances of any living members of this species still existing in the hinterlands of Flores are extremely remote, but some observers have noted that the indigenous peoples of Flores recount a myth of small hairy humans who descend from the highlands to steal food and supplies.

Where Did Modern Humans Evolve?

Young Earth creationists have a ready-made answer to this question—the Garden of Eden—and they accept the biblical story of Adam and Eve as factual history instead of mythic saga. IDers, by comparison, avoid all such biblical references and argue simply that there is no evidence for human ancestry and thus there must have been a divine spark of humanity miraculously inserted into one hominid species. The evolutionary story is more complex, says

University of Cambridge professor Peter Forster, an expert in archaeogenetics, who demonstrated how prehistoric human migrations can be traced by mitochondrial DNA (mtDNA) through the maternal line of modern humans.

Forster outlined our migrational history over the past 200,000 years as follows: Between 190,000 and 130,000 years ago, a single female, known formally as the "mitochondrial coalescent" but dubbed "mitochondrial Eve," gave rise to every living human today. Between 80,000 and 60,000 years ago, a large population from the center of Africa migrated to all areas of Africa, as well as the area of present-day Saudi Arabia. This migration may have taken two routes, a northern one up the Nile and around the Red Sea and a southern one across the narrow strait at Yemen which during the last ice age would have been only five kilometers across (Forster thinks the southern route was the more likely). Between 60,000 and 30,000 years ago there was a great migration to Southeast Asia, Northern Asia, and Europe. Between 30,000 and 20,000 years ago, people spread throughout most of the rest of the world, including Australia, and between 20,000 and 15,000 years ago they migrated into North America, making their way into South America between 15,000 and 2,000 years ago. The final migration over the past 2,000 years saw the settlement of the Pacific islands.

Creationists who take comfort in the "Eve" part of this account should know that other scientists believe that we arose from a single population, not one individual, and still other scientists opt for the multiregional theory of human origins, in which different human groups arose from different ancestral groups who migrated out of Africa at different times. And in any case, this is just the story of our most recent ancestors; our lineage goes back at least as far as six million years ago to the common ancestor of apes and humans.

What Direct Evidence Is There
for Natural Selection and Evolution?

Every year for the past three decades the husband and wife team of Peter and Rosemary Grant, both from Princeton University, have parked themselves on the Galápagos island of Daphne Major, a tiny volcanic plug 120 meters high and a kilometer long, to study the evolution of Darwin's finches.

We now have extensive evidence from both fossils and genes that three million years ago a single group of finches flew out to the Galápagos during a time of very active plate tectonics and the creation of the island archipelago. When this founder population arrived, it encountered a permanent El Niño that made the islands warm and wet, which helped to trigger an explosion of speciation.

First came the warbler finch, then the tree finch (of which there are now five species), and then the ground finch (of which there are now six species). Following Ernst Mayr's theory of allopatric speciation (in which a founder daughter population breaks away from the parental population), the first finches landed on San Cristóbal, then migrated to Española, then to Floreana, then to Santa Cruz, and finally made their way back to San Cristóbal. Along the journey the finches adapted to local conditions. Finches in highlands developed larger beaks to break hard beetles and seeds. Finches in lowlands evolved smaller beaks for eating small seeds and succulents. As an opportunistic species, some of these finches also ate sea turtle eggs and sucked the blood from blue-footed boobies. Different adaptations to different islands led to the origin of new species.

The Grants provide one of the strongest examples of observing evolution as it unfolds, which they did in tracking thirty years of environmental changes on Daphne Major and how the finch species responded. Arriving in 1973, the Grants immediately witnessed a drought that wiped out 85 percent of the population of two species of finches (the ground finch *Geospiza fortis* and the cactus finch *Geospiza scandens*). From 1975 to 1978 there was almost no rainfall and natural selection operated rapidly to change beak size. In 1983, an El Niño rainfall produced an abundance of plants and trees and cactus fruit, but two years later the island dried out and the large seeds were replaced by small seeds. Throughout this cycle, beak shape, beak size, and body size of the finches all changed in parallel.

What Is the Difference between
Natural Selection and Sexual Selection?

On the heels of Margulis's pronouncement of the death of neo-Darwinism, Stanford University evolutionary biologist Joan Roughgarden proclaimed the demise of Darwin's theory of sexual selection. Darwin said that males have stronger passions than females, that females are coy, and that females choose mates who are more attractive, vigorous, and well-armed. "People are surprised to learn how much sex animals have for purely social reasons (including same-sex sexuality in over 300 species of vertebrates)," Roughgarden explained, "and how many species have sex-role reversal in which the males are drab and the females are colorfully ornamented and compete for the attention of males, and that most plants and perhaps a quarter of all animal species have individuals that cannot be classified as male or female."

In response to Roughgarden, University of Georgia evolutionary biologist Patricia Gowaty noted that Roughgarden is right in identifying the exceptions to Darwin's theory and that there is much we still do not know, but added that since Darwin's time much has been learned about mate selection and competition that should not be dismissed.

What Is Right in Evolutionary Theory?

At the end of the Evolution Summit the evolutionary biologist Douglas Futuyma, who wrote the book on evolution (literally—he is the author of the bestselling textbook on evolution biology), summed up the state of evolutionary theory today: "I am tempted to quote from Gilbert and Sullivan's *The Mikado*: 'I am right and you are right and all is right as right can be.'" Futuyma explained that he had agreements with everyone on some aspects of the various debates and controversies under discussion, but that in the end more research and more data will resolve some issues and open up new ones.

Here are a few additional controversies that are well within the borders of mainstream evolutionary science and that are experiencing vigorous debate:

1. If natural selection is the primary mechanism of evolution, what is the role of chance and contingency in the history of life?
2. What was the origin of organic molecules?
3. If DNA came from RNA, what was the first replicator that became RNA?
4. What is the target of natural selection? (Strict Darwinians believe that the individual organism is the sole target of selection; others hold that selection may occur below the individual at

the level of genes, chromosomes, organelles, and cells, and above the individual at the level of groups, species, and multi-species communities.)

5. What is the relationship between evolution and embryological development (evo-devo)?

6. How much of modern human behavior can be explained by our evolutionary history (evolutionary psychology v. learning psychology; nature v. nurture)?

7. How much of modern society, culture, politics, and economics can be explained by our evolutionary past (evolutionary economics, Darwinian politics)?

Science's greatest strength lies in the ability not only to withstand such buffeting, but actually to grow from it. Creationists, IDers, and outsiders contend that science is a cozy and insular club in which meetings are held to enforce agreement with the party line, to circle the wagons against any and all would-be challengers, and to achieve consensus on the most contentious issues. This conclusion cannot have been proposed by anyone who has ever attended a scientific conference. The World Summit on Evolution, like most scientific conferences, revealed a science rich in controversy and debate as well as data and theory. After a century and a half of such disputation, the theory of evolution has never been stronger.

EPILOGUE

✦

Why Science Matters

I worry that, especially as the Millennium edges nearer, pseudoscience and superstition will seem year by year more tempting, the siren song of unreason more sonorous and attractive. Where have we heard it before? Whenever our ethnic or national prejudices are aroused, in times of scarcity, during challenges to national self-esteem or nerve, when we agonize about our diminished cosmic place and purpose, or when fanaticism is bubbling up around us—then, habits of thought familiar from ages past reach for the controls. The candle flame gutters. Its little pool of light trembles. Darkness gathers. The demons begin to stir.

—Carl Sagan, *The Demon-Haunted World*, 1996

Many millennia ago, the Esselen Indians of the California coast frequented natural hot springs just south of what later Spanish explorers would name Monterey Bay. The near-boiling waters of the hot springs cascaded out of the cliffs and into the crashing waves of the Pacific Ocean below. The Esselen found the sulfur-rich waters relaxing, the morning mist and afternoon sun rejuvenating, and the spectacular views of mountains and beaches breathtaking. It was a spiritual center, a place to go to renew one's soul.

In 1910, with the Esselen Indians long ago extirpated by European guns, germs, and steel, Dr. Henry Murphy purchased the land and constructed tubs to capture the hot springs for the restoration of his patients' health. In 1962, Dr. Murphy's grandson, Michael Murphy, and an associate named Richard Price transformed the site into a center for the nascent human potential movement, calling it the Esalen Institute in honor of the original residents. Today, Esalen is a cluster of meeting rooms, lodging facilities, and architecturally elegant hot baths all nestled into a stunning craggy outcrop along the Pacific Coast Highway.

Over the decades Esalen has hosted a veritable Who's Who of savants and gurus, including Alan Watts, Aldous Huxley, Abraham Maslow, Paul Tillich, Arnold Toynbee, B. F. Skinner, Stanislav Grof, Ida Rolf, Carl Rogers, Linus Pauling, Buckminster Fuller, Rollo May, Joseph Campbell, Susan Sontag, Ken Kesey, Gregory Bateson, John C. Lilly, Carlos Castaneda, Fritjof Capra, Ansel Adams, John Cage, Joan Baez, Robert Anton Wilson, Andrew Weil, Deepak Chopra, and even counterculture icon Bob Dylan. I had long wanted to visit Esalen, ever since I read *Surely You're Joking, Mr. Feynman,* in which the Nobel laureate Caltech physicist Richard Feynman recounted his experiences in the natural hot spring baths there. In one particularly amusing tale, a woman was getting a massage from a man she just met: "He starts to rub her big toe. 'I think I feel it,' he says. 'I feel a kind of dent—is that the pituitary?' I blurt out, 'You're a helluva long way from the pituitary, man!' They looked at me horrified and said, 'It's reflexology!' I quickly closed my eyes and appeared to be meditating."[1]

With this background to the mecca of the New Age movement, I was delighted to receive an invitation to speak at a conference there on evolutionary theory that featured an eclectic mix of participants: an anthropologist, a philosopher of religion, a Buddhist monk, a biophysicist, a philosopher of mind, an evolutionary biologist, a psychologist, a complexity theorist, a business entrepreneur, and a skeptic. The lectures and discussions were wide-ranging and diverse, but focused on the philosophical, religious, social, and spiritual implications and applications of evolution. My talk on the evolutionary origins of morality led to an invitation to teach a weekend seminar on science and spirituality at Esalen the following summer. Given my propensity for skepticism when it comes to most of the paranormal piffle proffered by the prajna peddlers meditating and soaking their way to nirvana at Esalen, I was surprised that the hall was full. Perhaps skeptical consciousness is rising!

The workshop was enriching for all of us, but it was in the extracurricular conversations—during healthful homegrown meals served cafeteria style with informal group seating, and while soaking in the hot tubs—that I gleaned a sense of what people believe and why. Once it became known that Mr. Skeptic was there, for example, I heard one after another "how do you explain *this*?" story, mostly involving angels, aliens, and the usual paranormal fare. But this being Esalen—ground zero for all that is weird and wonderful in the human potential movement—there were some singularly unusual accounts.

One woman explained the theory behind "bodywork," a combination of massage and "energy work" that involves adjusting the body's seven energy centers called chakras. I signed up for a massage, which was the best I've ever had, but when another practitioner told me about how she cured a woman's migraine headache by directing a light beam through her head, I decided that practice and theory are best kept separate. Another woman warned about the epidemic of satanic cults throughout Europe and America. "But there's no evidence of such cults," I countered. "Of course not," she explained. "They erase all memories and evidence of their nefarious activities." Of course.

One gentleman recounted a lengthy tantric sexual encounter with his lover that lasted for many hours, at the culmination of which a lightning bolt shot through her left eye followed by a blue light-being child entering her womb, ensuring conception. Nine months later, friends and gurus joined the couple in a hot house, sweating their way through their own "rebirthing" process (to cleanse the pain of one's own childbirth so that it is not passed on to the child) before the mother gave birth to a baby boy. Right then and there the father informed this infant that he would need to become an athlete in order to get into college; two decades later, the father told me as I slipped deeper into the hot tub, this young man became a professional baseball player. "How do you explain *that*?" he queried. I quickly closed my eyes and appeared to be meditating.

People have and share such spiritual experiences, and impart larger significance to them, because we have a cortex large enough to conceive of such transcendent notions, and an imagination creative enough to concoct fantastic narratives. If we define the spirit (or soul) as the pattern of information of which we are made—our genes, proteins, memories, and personalities—then spirituality is the quest to know the place of our essence within the deep time of evolution and the deep space of the cosmos.

There are many ways to be spiritual, and science is one in its awe-inspiring account of who we are and where we came from. "The universe is all that is, or ever was, or ever will be," began the late astronomer Carl Sagan in the opening scene of *Cosmos,* filmed just down the coast from Esalen. "Our contemplations of the cosmos stir us. There's a tingling in the spine, a catch in the voice, a faint sensation as if a distant memory of falling from a great height.

We know we are approaching the grandest of mysteries." How can we connect to this vast cosmos? Sagan's answer is both spiritually scientific and scientifically spiritual: "The cosmos is within us. We are made of star stuff," he said, referring to the stellar origins of the chemical elements of life, which are cooked in the interiors of stars, then released in supernova explosions into interstellar space where they condense into a new solar system with planets, some of which have life that is composed of this star stuff. "We've begun at last to wonder about our origins, star stuff contemplating the stars, organized collections of ten billion billion billion atoms contemplating the evolution of matter, tracing that long path by which it arrived at consciousness here on the planet Earth and perhaps throughout the cosmos. Our obligation to survive and flourish is owed not just to ourselves but also to that cosmos, ancient and vast, from which we spring."[2]

That is spiritual gold, and Carl Sagan was one of the most spiritual scientists of our epoch.[3]

How can we find spiritual meaning in a scientific worldview? Spirituality is a way of being in the world, a sense of one's place in the cosmos, a relationship to that which extends beyond oneself. There are many sources of spirituality. Unfortunately, there are those who believe that science and spirituality are in conflict. The nineteenth-century English poet John Keats, for example, lamented that Isaac Newton had "destroyed the poetry of the rainbow by reducing it to a prism." Natural philosophy, he complained in his 1820 poem *Lamia*,

> *will clip an Angel's wings,*
> *Conquer all mysteries by rule and line,*
> *Empty the haunted air, and gnomed mine—*
> *Unweave a rainbow.*

Keats's contemporary Samuel Taylor Coleridge similarly averred, "the souls of 500 Sir Isaac Newtons would go to the making up of a Shakespeare or a Milton."[4]

Does a scientific explanation for the world diminish its spiritual beauty? I think not. Science and spirituality are complementary, not conflicting; additive, not detractive. Anything that generates a sense of awe may be a source of spirituality. Science does this in spades. I am deeply moved, for example, when I observe through my Meade eight-inch reflecting telescope in my backyard the fuzzy little patch of light that is the Andromeda galaxy. It is not just because it is lovely, but because I also understand that the photons of light landing on my retina left Andromeda 2.9 million years ago, when our ancestors were tiny-brained hominids roaming the plains of Africa.

I am doubly stirred because it was not until 1923 that the astronomer Edwin Hubble, using the 100-inch telescope on Mount Wilson just above my home in the foothills of Pasadena, discovered that this "nebula" was actually an extragalactic stellar system of immense size and distance. Hubble subsequently discovered that the light from most galaxies is shifted toward the red end of the electromagnetic spectrum (literally unweaving a rainbow of colors), meaning that the universe is expanding away from an explosive creation. It was the first empirical evidence indicating that the universe had a beginning, and thus is not eternal. What could be more awe-inspiring—more numinous, magical, spiritual—than this cosmic visage? Mount Wilson Observatory is the Chartres Cathedral of our time.

Since I live in Southern California, I have had many occasions to make the climb to Mount Wilson, a twenty-five-mile trek from the bedroom community of La Cañada up a twisting mountain road whose terminus is a cluster of telescopes, interferometers, and communications towers that feed the mega-media conglomerate below. As a young student of science in the 1970s, I took a general tour. As a serious bicycle racer in the 1980s, I rode there every Wednesday (a tradition still practiced by a handful of us cycling diehards). In the 1990s, I took several scientists there, including the late Harvard evolutionary theorist Stephen Jay Gould, who described it as a deeply moving experience. Most recently, in November of 2004, I arranged a visit to the observatory for the British evolutionary biologist Richard Dawkins. As we were standing beneath the magnificent dome housing the 100-inch telescope and pondering how marvelous, even miraculous, this scientistic vision of the cosmos and our place in it all seemed, Dawkins turned to me and said, "All of this makes me so proud of our species that I am almost moved to tears."

As we are pattern-seeking, story-telling primates, to most of us the pattern of life and the universe indicates design. For countless millennia we have taken these patterns and constructed stories about how life and the cosmos were designed specifically for us from above. For the past few centuries, however, science has presented us with a viable alternative in which the design comes from below through the direction of built-in self-organizing principles of emergence and complexity. Perhaps this natural process, like the other natural forces which we are all comfortable accepting as nonthreatening to religion, was God's way of creating life. Maybe God *is* the laws of nature—or even nature itself—but this is a theological supposition, not a scientific one.

What science tells us is that we are but one among hundreds of millions of species that evolved over the course of three and a half billion years on one tiny planet among many orbiting an ordinary star, itself one of possibly billions of solar systems in an ordinary galaxy that contains hundreds of billions of stars, itself located in a cluster of galaxies not so different from millions of other galaxy clusters, themselves whirling away from one another in an expanding cosmic bubble universe that very possibly is only one among a near infinite number of bubble universes. Is it really possible that this entire cosmological multiverse was designed and exists for one tiny subgroup of a single species on one planet in a lone galaxy in that solitary bubble universe? It seems unlikely.

Herein lies the spiritual side of science—*sciensuality*, if you will pardon an awkward neologism but one that echoes the sensuality of discovery. If religion and spirituality are supposed to generate awe and humility in the face of the creator, what could be more awesome and humbling than the deep space discovered by Hubble and the cosmologists, and the deep time discovered by Darwin and the evolutionists?

Darwin matters because evolution matters. Evolution matters because science matters. Science matters because it is the preeminent story of our age, an epic saga about who we are, where we came from, and where we are going.

CODA

✦

Genesis Revisited

The fundamental difference between evolutionary theory and Intelligent Design is the nature of explanation: natural versus supernatural. The problem with the supernatural explanations of Intelligent Design is that there is nothing we can *do* with supernatural explanations. They lead to no data collection, no testable hypotheses, no quantifiable theories: therefore, no science.

To demonstrate the logical absurdity of trying to squeeze the round peg of science into the square hole of religion, I offer the following scientific revision of the Genesis creation story. This is not intended as a sacrilege of the mythic grandeur of Genesis; rather, it is a mere extension of what the creationists have already done to Genesis in their insistence that it be read not as mythological saga but as scientific prose. If Genesis were written in the language of modern science, it would read something like this:

In the beginning—specifically on October 23, 4004 BC, at noon— out of quantum foam fluctuation God created the Big Bang, followed by cosmological inflation and an expanding universe. And darkness was upon the face of the deep, so He created Quarks and therefrom He created hydrogen atoms and thence He commanded the hydrogen atoms to fuse and become helium atoms and in the process to release energy in the form of light. And the light maker He called the sun, and the process He called fusion. And He saw the light was good because now He could see what he was doing, so he created Earth. And the evening and the morning were the first day.

And God said, Let there be lots of fusion light makers in the sky. Some of these fusion makers He grouped into collections He called galaxies, and these appeared to be millions and even billions of light-years from Earth, which would mean that they were created before the first creation in 4004 BC. This was confusing, so God created tired light, and the creation story was preserved. And created He many wondrous splendors such as Red Giants, White Dwarfs, Quasars, Pulsars, Supernovas, Worm Holes, and even Black Holes out of which nothing can escape. But since God cannot be constrained by nothing, He created Hawking radiation through which information can escape from Black Holes. This made God even more tired than tired light, and the evening and the morning were the second day.

And God said, Let the waters under the heavens be gathered together unto one place, and let the continents drift apart by plate tectonics. He decreed that sea floor spreading would create zones of emergence, and He caused subduction zones to build mountains and cause earthquakes. In weak points in the crust God created volcanic islands, where the next day He would place organisms that were similar to but different from their relatives on the continents, so that still later created creatures called humans would mistake them for evolved descendants created by adaptive radiation. And the evening and the morning were the third day.

And God saw that the land was barren, so He created animals

bearing their own kind, declaring Thou shalt not evolve into new species, and thy equilibrium shall not be punctuated. And God placed into the rocks, fossils that appeared older than 4004 BC that were similar to but different from living creatures. And the sequence resembled descent with modification. And the evening and the morning were the fourth day.

And God said, Let the waters bring forth abundantly the moving creatures that have life, the fishes. And God created great whales whose skeletal structure and physiology were homologous with the land mammals He would create later that day. God then brought forth abundantly all creatures, great and small, declaring that microevolution was permitted, but not macroevolution. And God said, "Natura non facit saltum"—Nature shall not make leaps. And the evening and the morning were the fifth day.

And God created the pongids and hominids with 98 percent genetic similarity, naming two of them Adam and Eve. In the book in which God explained how He did all this, the Bible, in one chapter He said He created Adam and Eve together out of the dust at the same time, but in another chapter He said He created Adam first, then later created Eve out of one of Adam's ribs. This caused confusion in the valley of the shadow of doubt, so God created theologians to sort it out.

And in the ground placed He in abundance teeth, jaws, skulls, and pelvises of transitional fossils from pre-Adamite creatures. One chosen as his special creation He named Lucy, who could walk upright like a human but had a small brain like an ape. And God realized this too was confusing, so he created paleoanthropologists to figure it out.

Just as He was finishing up the loose ends of the creation, God realized that Adam's immediate descendants would not understand inflationary cosmology, global general relativity, quantum mechanics, astrophysics, biochemistry, paleontology, and evolutionary biology, so he created creation myths. But there were so many creation stories throughout the world that God realized this too was confusing, so created He anthropologists and mythologists to explain all that.

By now the valley of the shadow of doubt was overrun with skepticism, so God became angry—so angry that God lost His temper and cursed the first humans, telling them to go forth and multiply themselves (but not in those words). But the humans took God literally and now there are over six billion of them. And the evening and the morning were the sixth day.

By now God was tired, so He proclaimed, "Thank Me it's Friday," and He made the weekend. It was a good idea.

APPENDIX

✦

Equal Time for Whom?

Over the past century, nearly every court case and curriculum dispute in the evolution-creation debate has included some form of the "equal time" argument. Well, even if we all agreed public school science classes should spend equal time on each perspective, we must ask, Equal time for whom? My friend and colleague Eugenie Scott, Executive Director of the National Center for Science Education, outlines at least eight different positions one might take on the creation-evolution continuum.[*] These include:

> *Young Earth Creationists,* who believe that the earth and all life on it were created within the last ten thousand years.

> *Old Earth Creationists,* who believe that the earth is ancient and that although microevolution may alter organisms into different varieties of species, all life was created by God and species cannot evolve into new species.

---

[*] For a fuller explication, visit their Web site at http://www.natcenscied.org/.

Gap Creationists, who believe that there was a large temporal gap between Genesis 1:1 and 1:2, in which a pre-Adam creation was destroyed and God recreated the world in six days; the time gap between the two separate creations allows for an accommodation of an old Earth with the special creation.

Day-Age Creationists, who believe that each of the six days of creation represents a geological epoch, and that the Genesis sequence of creation roughly parallels the sequence of evolution.

Progressive Creationists, who accept most scientific findings about the age of the universe and that God created "kinds" of animals sequentially; the fossil record is an accurate representation of history because different animals and plants appeared at different times rather than having been created all at once.

Intelligent Design Creationists, who believe that the order, purpose, and design found in the world are proof of an Intelligent Designer.

Evolutionary Creationists, who believe that God used evolution to bring about life according to his foreordained plan from the beginning.

Theistic Evolutionists, who believe that God used evolution to bring about life, but intervenes at critical intervals during the history of life.

Note that the Intelligent Design creationists are but one of many competing for space in the curriculum; if the government were to force teachers to grant equal time for them, then why not these others? And this short list does not include the creation theories of other cultures, such as:

No Creation Story from India, where the world has always existed as it is now, unchanging from eternity.

The Slain Monster Creation Story from Sumeria-Babylonia, in which the world was created from the parts of a slain monster.

The Primordial Parents Creation Stories from the Zuñi Indians, Cook Islanders, and Egyptians, in which the world was created by the interaction of primordial parents.

The Cosmic Egg Creation Stories from Japan, Samoa, Persia, and China, in which the world was generated from an egg.

The Spoken Edict Creation Stories from the Mayans, the Egyptians, and the Hebrews, in which the world sprang into being at the command of a god (this is the belief of creationists and Intelligent Design theorists).

The Sea Creation Stories from the Burmese, Choctaw Indians, and Icelanders, in which the world was created from out of the sea.

If equal time were given to all of these positions, along with the many other creation myths from diverse cultures around the world, when would students have time for science?

NOTES

✦

Prologue: Why Evolution Matters

1. We were accompanied on this expedition by botanist Phil Pack, snail specialist Robert Smith, explorer Daniel Bennett, and medical engineer Chuck Lemme.

2. Sulloway's historical reconstruction of the development of Darwin's evolutionary thinking can be found in a number of his papers: Frank Sulloway, "Darwin and His Finches: The Evolution of a Legend," *Journal of the History of Biology* 15 (1982), pp. 1–53; "Darwin's Conversion: The *Beagle* Voyage and Its Aftermath," *Journal of the History of Biology* 15 (1982), pp. 325–96; "The Legend of Darwin's Finches," *Nature* 303 (1983), p. 372; "Darwin and the Galapagos," *Biological Journal of the Linnean Society* 21 (1984), pp. 29–59.

3. Letter to Joseph Hooker dated January 14, 1844, quoted in Janet Browne, *Voyaging: Charles Darwin. A Biography* (New York: Knopf, 1995), p. 452.

4. Darwin would have waited even longer had he not rushed into print for priority's sake because the naturalist Alfred Russel Wallace had sent Darwin his own theory of evolution the year before. For a detailed account of the "priority dispute" between Darwin and Wallace, see Michael Shermer, *In Darwin's Shadow: The Life and Science of Alfred Russel Wallace* (New York: Oxford University Press, 2002).

5. Ernst Mayr, *Growth of Biological Thought* (Cambridge, Mass.: Harvard University Press, 1982), p. 495.

6. All quotes on the reaction to Darwin's theory are from K. Korey, *The Essential Darwin: Selections and Commentary* (Boston, Mass.: Little, Brown, 1984).

7. Interestingly, a sizable 41 percent believe that parents, rather than scientists (28 percent) or school boards (21 percent), should be responsible for teaching

children about the origin and evolution of life. Pew Research Center for the People & the Press survey data available online at http://people-press.org/reports/display.php3?ReportID=254.

8. Elisabeth Bumiller, "Bush Remarks Roil Debate on Teaching of Evolution," *New York Times*, August 3, 2005.

9. I have written about this at length in my book *How We Believe: Science, Skepticism, and the Search for God* (New York: Times Books, 1999).

10. Adapted and paraphrased from Ernst Mayr, *The Growth of Biological Thought* (Cambridge, Mass.: Harvard University Press, 1982), p. 501.

11. Theodosius Dobzhansky, "Nothing in Biology Makes Sense Except in the Light of Evolution," *American Biology Teacher* 35 (1973), pp. 125–29.

1. The Facts of Evolution

1. Letter reprinted in Francis Darwin, *The Life and Letters of Charles Darwin*, Vol. 2 (London: John Murray, 1887), p. 121.

2. When Darwin was in college there was a debate raging over the concept of induction—what it is and how it is used in science. Although definitions varied, it was roughly understood to mean arguing from the specific to the general, from data to theory. In 1830, the astronomer John Herschel argued that induction was reasoning from the known to the unknown. In 1840, the philosopher of science William Whewell insisted that induction was the superimposing of concepts on facts by the mind, even if they are not empirically verifiable. In 1843, the philosopher John Stuart Mill claimed that induction was the discovery of general laws from specific facts, but that they had to be verified empirically. Kepler's discovery of the laws of planetary motion were a classic case study of induction. For Herschel and Mill, Kepler discovered these laws through careful observation and induction. For Whewell, the laws were self-evident truths that could have been known a priori. By the 1860s, as the theory of evolution was gaining momentum and converts, Herschel and Mill carried the day, not so much because they were right and Whewell was wrong, but because empiricism was becoming integral to the understanding of how good science is done. This drove Darwin to compile copious data for his theory before going public. Classic texts in this debate include John F. W. Herschel, *Preliminary Discourse on the Study of Natural Philosophy* (London: Longmans, Rees, Orme, Brown and Green, 1830); William Whewell, *The Philosophy of the Inductive Sciences* (London: J. W. Parker, 1840); and John Stuart Mill, *A System of Logic, Ratiocinative and Inductive, Being a Connected View of the Principles of Evidence, and the Methods of Scientific Investigation* (London: Longmans, Green, 1843).

3. Francis Darwin (ed.), *The Autobiography of Charles Darwin and Selected Letters* (New York: Dover Publications, 1958), p. 98. Originally published 1892.

4. T. H. Huxley, *Darwiniana* (New York: Appleton, 1896), p. 72.

5. In Francis Darwin, *More Letters of Charles Darwin*, Vol. 2 (London: John Murray, 1903), p. 323.

6. Francis Darwin (ed.), *Autobiography of Charles Darwin*.

7. John Ray, *The Wisdom of God Manifested in Works of the Creation* (London: Samuel Smith, 1691).

8. William Paley, *Natural Theology: or, Evidences of the Existence and Attributes of the Deity, Collected from the Appearances of Nature* (London: E. Paulder, 1802).

9. For a thorough discussion of Paley's influence on Darwin, see Keith Thomson, *Before Darwin* (New Haven, Conn.: Yale University Press, 2005).

10. Letter from Charles Darwin to John Lubbock, November 15, 1859, in Francis Darwin, *The Life and Letters of Charles Darwin*, Vol. 2, p. 8.

11. Ernst Mayr, *Toward a New Philosophy of Biology* (Cambridge, Mass.: Harvard University Press, 1988).

12. Ernst Mayr, "Species Concepts and Definitions," in *The Species Problem* (Washington, D.C.: American Association for the Advancement of Science Publication 50, 1957). Mayr offers this expanded definition: "A species consists of a group of populations which replace each other geographically or ecologically and of which the neighboring ones intergrade or hybridize wherever they are in contact or which are potentially capable of doing so (with one or more of the populations) in those cases where contact is prevented by geographical or ecological barriers." See also Ernst Mayr, *Evolution and the Diversity of Life* (Cambridge, Mass.: Harvard University Press, 1976).

13. Charles Darwin, *On the Origin of Species by Means of Natural Selection: or, The Preservation of Favoured Races in the Struggle for Life* (London: John Murray, 1859), p. 63.

14. Charles Darwin, *Origin of Species*, p. 84.

15. Richard Dawkins, *The Selfish Gene* (Oxford: Oxford University Press, 1976).

16. Percival W. Davis and Dean H. Kenyon, *Of Pandas and People* (Dallas, Tex.: Haughton, 1993).

17. Charles Darwin, *Origin of Species*, p. 280.

18. For a brief analysis of whether punctuated equilibrium constitutes a paradigm shift in evolutionary theory, see my book *The Borderlands of Science* (New York: Oxford University Press, 2001). For a lengthy summary of all critiques of punctuated equilibrium, and detailed responses to them, see the 300-page chapter on the subject in Stephen Jay Gould, *The Structure of Evolutionary Theory* (Cambridge, Mass.: Harvard University Press, 2002).

19. Niles Eldredge and Stephen Jay Gould, "Punctuated Equilibria: An Alternative to Phyletic Gradualism," in T. J. M. Schopf (ed.), *Models in Paleobiology* (San Francisco: Freeman, 1972), p. 205.

20. D. S. McKay et al., "Search for Past Life on Mars: Possible Relic Biogenic Activity in Martian Meteorite ALH84001," *Science* 273 (1996), pp. 924–30.

21. William Schopf, *Cradle of Life: The Discovery of Earth's Earliest Fossils* (Princeton, N.J.: Princeton University Press, 1999).

22. Whewell, *Philosophy of the Inductive Sciences*, p. 230. The irony is that the theory of evolution is arguably the most consilient theory ever generated, and Whewell rejected it, going so far as to block the *Origin of Species* from being shelved at the library at Trinity College, Cambridge.

23. Personal correspondence, December 13, 2004.

24. All three articles appear in the November 22, 2002, issue of *Science:* Jennifer A. Leonard et al., "Ancient DNA Evidence for Old World Origin of New World Dogs," pp. 1613–16; Peter Savolainen et al., "Genetic Evidence for an East

Asian Origin of Domestic Dogs," pp. 1610–13; Brian Hare et al., "The Domestication of Social Cognition in Dogs," pp. 1634–36.

25. Richard Dawkins, *The Ancestor's Tale: A Pilgrimage to the Dawn of Evolution* (Boston: Houghton Mifflin, 2004).

26. Luigi Luca Cavalli-Sforza, P. Menozzi, and A. Piazza, *The History and Geography of Human Genes* (Princeton, N.J.: Princeton University Press, 1994).

27. Jack Horner, *Digging Dinosaurs* (New York: Harper & Row, 1988), p. 168.

28. Ibid., p. 129.

29. Ibid., pp. 129–43.

2. Why People Do Not Accept Evolution

1. All quotes are from *Bryan's Last Speech: The Most Powerful Argument against Evolution Ever Made,* a small booklet (price 25¢) published shortly after his death and reprinted in full in *Skeptic* Vol. 4, No. 2 (1998), pp. 88–100. The booklet was sent to me by my friend and colleague Clayton Drees, who found it in a used book store in Virginia. In the film about the trial, *Inherit the Wind,* in the middle of Bryan's final moving speech he dramatically keels over dead in the courtroom to the gasps of his faithful followers and the chagrin of his evolutionary opponents. The reality was perhaps a bit less histrionic, but the real speech is much more revealing (in the film he is reduced to reciting the books of the Bible).

2. I discuss the Scopes trial briefly in both *Why People Believe Weird Things •* and *How We Believe.* For a complete history of the trial see Edward J. Larson, *Summer for the Gods: The Scopes Trial and America's Continuing Debate over Science and Religion* (New York: Basic Books, 1997).

3. Quotes in this section are from Stephen Jay Gould, "William Jennings Bryan's Last Campaign," *Natural History* (November 1987), pp. 32–38.

4. J. V. Grabiner and P. D. Miller, "Effects of the Scopes Trial," *Science* 185 (1974), pp. 832–36.

5. V. L. Kellog, *Headquarters Nights* (Boston: The Atlantic Monthly Press. 1917). Kellog had joined the Belgian relief program before America entered the war and, through his contacts with professional scientists, gained access to the German General Staff in Berlin, from whom he gathered the material for his book. Gould's 1987 reconstruction of Bryan's intellectual reversal on the theory of evolution is unsurpassed (see note 3 above).

6. Quoted in T. C. Riniolo, "The Attorney and the Shrink: Clarence Darrow, Sigmund Freud, and the Leopold and Loeb Trial," *Skeptic* Vol. 9, No. 3 (2002), pp. 44–48.

7. Darrow's defense echoes that used by the Menendez brothers' attorney Leslie Abrams decades later, when she tried to get the boys off from the murder of their parents by arguing they were the victims of parental abuse.

8. Historian Richard Hofstadter called this "probably the most effective speech in the history of American party politics."

9. Thomas H. Huxley, "The Origin of Species" (review), *Westminster Review* 17 (1860), pp. 541–70. Ernst Mayr, *Toward a New Philosophy of Biology* (Cambridge, Mass.: Harvard University Press, 1988), p. 161. Stephen Jay Gould, *The*

Structure of Evolutionary Theory (Cambridge, Mass.: Harvard University Press, 2002). Richard Dawkins, *A Devil's Chaplain* (New York: Houghton Mifflin, 2003), p. 78.

10. Quoted in R. Bailey, "Origin of the Specious," *Reason* (July 1997).

11. The three-hour briefing was held on May 10, 2000. Quoted in D. Wald, "Intelligent Design Meets Congressional Designers," *Skeptic* Vol. 8, No. 2 (2002), pp. 16–17.

12. For a thorough discussion on the liberal resistance to evolutionary theory, particularly when applied to human behavior, see Steven Pinker, *The Blank Slate: The Modern Denial of Human Nature* (New York: Viking, 2002).

13. *Evolution denial* is the doppelganger of *Holocaust denial,* in that evolution deniers use techniques of rhetoric and debate similar to those of Holocaust deniers. For a full discussion see my book *Denying History* (Berkeley: University of California Press, 2000). See also Massimo Pigliucci, *Denying Evolution* (Sunderland, Mass.: Sinauer, 2002).

14. Eugenie Scott and the National Center for Science Education track hundreds of specific incidents of teachers' silence in the face of controversy. See www.ncse.org.

3. In Search of the Designer

1. Michael Shermer and Frank. J. Sulloway, "Religion and Belief in God: An Empirical Study," in press 2006. We received a total of 1,002 responses out of 10,000 surveys obtained from Survey Sampling, Inc., Fairfield, Connecticut. The average age of respondents was 42.2 ($SD = 15.9$). Correlations and significance values were: being raised in a religious manner ($r = .39$, $N = 985$, $t = 13.23$, $p<.0001$), parents' religiosity ($r = .29$, $N = 984$, $t = 9.63$, $p<.0001$), lower levels of education ($r = -.21$, $N = 977$, $t = -6.67$, $p<.0001$), gender (women are more religious than men, $r = .15$, $N = 980$, $t = 4.90$, $p<.0001$), coming from a large family ($r = .12$, $N = 878$, $p<.001$), conflict with parents ($r = -.09$, $N = 959$, $t = -2.66$, $p<.01$), and age ($r = -.06$, $N = 976$, $t = -1.80$, $p<.08$).

2. S. A. Vyse, *Believing in Magic: The Psychology of Superstition* (New York: Oxford University Press, 1997), pp. 84–85.

3. Clarke's laws are available online at http://www.lsi.usp.br/~rbianchi/clarke/ACC.Laws.html. Clarke's First Law: "When a distinguished but elderly scientist states that something is possible he is almost certainly right. When he states that something is impossible, he is very probably wrong." Clarke's Second Law: "The only way of discovering the limits of the possible is to venture a little way past them into the impossible." Clarke's First Law was first published in "Hazards of Prophecy: The Failure of Imagination," an essay in his 1962 book *Profiles of the Future*. The second law was originally a derivative of the first and it became "Clarke's Second Law" later, after Clarke proposed the Third Law in a revised 1973 edition of *Profiles of the Future* because, he said, "As three laws were good enough for Newton, I have modestly decided to stop there."

4. I first proposed *Shermer's Last Law* in my column "Shermer's Last Law," *Scientific American* (January 2002), p. 33. I do not believe in naming laws

after oneself, so as the good book warns: the last shall be first and the first shall be last.

5. *Voyager* spacecraft speed and distances are available online at http://vraptor.jpl.nasa.gov/flteam/weekly-rpts/current.html#RTLT.

6. See Ray Kurzweil, *The Age of Spiritual Machines: When Computers Exceed Human Intelligence* (New York: Penguin, 1999), and *The Singularity Is Near: When Humans Transcend Biology* (New York: Viking, 2005).

7. Unless we happen to be the first space-faring species, which the Copernican Principle (that we are not special) predicts is unlikely.

8. Langdon Gilkey, *Creationism on Trial: Evolution and God at Little Rock* (Minneapolis, Minn.: Winston Press, 1985).

9. Langdon Gilkey, *Maker of Heaven and Earth: A Study of the Christian Doctrine of Creation* (New York: Doubleday, 1965). I am grateful to Michael McGough's insightful essay on why Intelligent Design is bad theology: Michael McGough, "Bad Science, Bad Theology," *Los Angeles Times,* August 15, 2005, p. C12.

10. Cited in S. J. Grenz and R. E. Olson, *20th Century Theology: God and the World in a Transitional Age* (Exeter, U.K.: Paternoster Press, 1993), p. 124.

11. Quoted in G. H. Smith, *Atheism: The Case against God* (Buffalo, N.Y.: Prometheus, 1989), p. 34.

4. Debating Intelligent Design

1. John Stuart Mill, *On Liberty* (London: Longman, Roberts & Green, 1859). Published in the same portentous year as Darwin's *Origin of Species.*

2. As a secondary benefit, we can reinforce skeptics with additional intellectual firepower for use in their own debates with True Believers and Fence Sitters. And on a tertiary level, we can witness to both cohorts that skeptics and scientists are thoughtful, witty, and affable, and sans horns, rancor, and pathos. To wit, after my debate with Hovind I was handed several notes from Christians whose feedback led me to conclude that, at the very least, they were convinced that skeptics and scientists are not Satanists. Here are two:

> I am a believer of Creation. However, I wanted to tell you I respected your professionalism in your execution of what you had to say. I almost want to apologize on behalf of some Creationists present tonight.
>
> I cannot say that I agree with you, but I would like to thank you for your professional presentation, unlike your opposition.

If you think I exaggerate the perception of skeptics as Satanists, a note given to me after the debate, from "an Evangelist Christian—Born again," reiterated this fear: "I just want to tell you that we fight against a spiritual world and Satan will do anything to blind your eyes from the truth. I just ask you to consider this as a possibility! I will be praying for you!" A common question I get at such debates is: "Why did you give up your faith?" The question is asked out of genuine curiosity, but there is often a substrate implied in the voice and revealed in the eyes: "This couldn't happen to me, could it?" When I answer in the affirmative that, indeed, it could happen to anyone who is intellectually honest in their search for answers to life's most ponderous questions, I am sometimes accused of a false faith *ab initio:* "You were never really a Christian." How convenient, and cognitively bullet-proof. But tell that to my annoyed siblings and non-Christian friends, who tolerated my nonstop evangelizing for seven years. The sentiments were quite real.

3. David Hume, *An Enquiry Concerning Human Understanding* (Chicago: University of Chicago Press, 1952). Originally published in 1758. Emphasis added.

4. Herbert Spencer, *Essays Scientific, Political and Speculative* (London: Williams & Norgate, 1891).

5. I discovered the Fossil Fallacy not in my research on evolution deniers, but in my study of the Holocaust deniers, who demand "just one proof" of the veracity of the central tenets of the Shoah. For example, they ask: Where are the Zyklon-B gas pellet induction holes in the roof of the gas chamber in Krema II at Auschwitz-Birkenau? "No holes, no Holocaust," they claim, a slogan even emblazoned on a T-shirt worn by Holocaust deniers. We have since found these holes, but the fallacy is in assuming that the Holocaust is a single event that can be proven by a single piece of data. Just as the Holocaust was thousands of events that occurred in thousands of places and is proven (reconstructed) through thousands of historical facts, evolution is a process and historical sequence that is proven through thousands of bits of data from numerous fields of science that together give us a rich portrait of the history of life. See my book *Denying History* (Berkeley: University of California Press, 2000).

6. Donald R. Prothero, "The Fossils Say Yes," *Natural History* (November 2005), pp. 52–56.

7. Isaac Newton (Robert Maynard Hutchins, ed., Andrew Motte, trans.), *Mathematical Principles of Natural Philosophy* (Chicago: University of Chicago Press, 1952), p. 273. Originally published in 1789.

8. In his foreword to Niall Shanks's book *God, the Devil, and Darwin* (New York: Oxford University Press, 2004), Richard Dawkins poignantly spelled this out in a clever fictional dialogue between two scientists. "Imagine a fictional conversation between two scientists working on a hard problem, say A. L. Hodgkin and A. F. Huxley who, in real life, won the Nobel Prize for their brilliant model of the nerve impulse," Dawkins begins.

"I say, Huxley, this is a terribly difficult problem. I can't see how the nerve impulse works, can you?"

"No, Hodgkin, I can't, and these differential equations are fiendishly hard to solve. Why don't we just give up and say that the nerve impulse propagates by Nervous Energy?"

"Excellent idea, Huxley, let's write the Letter to *Nature* now, it'll only take one line, then we can turn to something easier."

9. For an extensive list of books by Intelligent Design creationists, and of books critical of Intelligent Design theory, see the bibliography.

10. Stephen Hawking, "Quantum Cosmology," in Stephen Hawking and Roger Penrose, *The Nature of Space and Time* (Princeton, N.J.: Princeton University Press, 1996), pp. 89–90.

11. John D. Barrow and Frank Tipler, *The Anthropic Cosmological Principle* (Oxford: Oxford University Press, 1988), p. vii.

12. Martin Rees, *Just Six Numbers: The Deep Forces That Shape the Universe* (New York: Basic Books, 2000).

13. Michael Denton, *Nature's Destiny: How the Laws of Biology Reveal Purpose in the Universe* (New York: Free Press, 1998).

14. John Barrow and John Webb, "Inconstant Constants," *Scientific American* (June 2005), pp. 57–63.

15. Raphael Bousso and Joseph Polchinski, "The String Theory Landscape," *Scientific American* (September 2004).

16. Victor Stenger, *The Unconscious Quantum: Metaphysics in Modern Physics and Cosmology* (Buffalo, N.Y.: Prometheus, 1995). Victor Stenger, "Is the Universe Fine-Tuned for Us?" in Matt Young and Taner Edis (eds.), *Why Intelligent Design Fails: A Scientific Critique of the New Creationism* (New Brunswick, N.J.: Rutgers University Press, 2004).

17. See Andrei Linde, *Particle Physics and Inflationary Cosmology* (New York: Academic Press, 1990); Quentin Smith, "A Natural Explanation of the Existence and Laws of Our Universe," *Australasian Journal of Philosophy* No. 68 (1990), pp. 22–43; Lee Smolin, *The Life of the Cosmos* (Oxford: Oxford University Press, 1997); and Alan Guth, *The Inflationary Universe: The Quest for a New Theory of Cosmic Origins* (Cambridge: Perseus Books, 1997). For an elegant summary of this field see James Gardner, *Biocosm* (Maui, Hawaii: Inner Ocean Publishing, 2003).

18. Stephen Hawking, "The Future of Theoretical Physics and Cosmology: Stephen Hawking 60th Birthday Symposium," Lecture at the Centre for Mathematical Sciences, Cambridge, United Kingdom, January 11, 2002.

19. Stephen C. Meyer, "Word Games: DNA, Design, and Intelligence," *Touchstone* Vol. 12, No. 4 (1999), pp. 44–50.

20. Voltaire quoted in B. R. Redman (ed.), *The Portable Voltaire* (New York: Penguin, 1985).

21. William Dembski, *No Free Lunch: Why Specified Complexity Cannot Be Purchased without Intelligence* (Lanham, Md.: Rowman & Littlefield, 2002).

22. William Dembski, *The Design Inference: Eliminating Chance through Small Probabilities* (New York: Cambridge University Press, 1998).

23. William Dembski, "The Intelligent Design Movement," *Cosmic Pursuit,* 1998. Available online at http://sapiensweb.free.fr/articles/2-dembski.htm.

24. Charles Darwin, *On the Origin of Species by Means of Natural Selection: or, The Preservation of Favoured Races in the Struggle for Life* (London: John Murray, 1859), p. 154.

25. Michael Behe, *Darwin's Black Box: The Biochemical Challenge to Evolution* (New York: Free Press, 1996), p. 39.

26. Ibid., pp. 232–33.

27. Michael Behe, "Molecular Machines: Experimental Support for the Design Inference," paper presented at the summer meeting of the C. S. Lewis Society, Cambridge University, United Kingdom, 1994. Available online at http://www.arn.org/docs/behe/mb_mm92496.htm.

28. Robert Pennock, *Tower of Babel: The Evidence against the New Creationism* (Cambridge, Mass.: MIT Press, 1999).

29. Jerry Coyne, "God in the Details," *Nature,* No. 383 (1996), pp. 227–28.

30. Charles Darwin, *On the Various Contrivances by Which British and Foreign Orchids Are Fertilized by Insects, and on the Good Effects of Intercrossing* (London: John Murray, 1862), p. 348.

31. Stephen Jay Gould and Elizabeth Vrba, "Exaptation: A Missing Term in the Science of Form," *Paleobiology* No. 8 (1982), pp. 4–15.

32. R. O. Prum and A. H. Brush, "Which Came First, the Feather or the Bird: A Long-Cherished View of How and Why Feathers Evolved Has Now Been Overturned," *Scientific American* (March 2003), pp. 84–93.

33. Kevin Padian and L. M. Chiappe, "The Origin of Birds and Their Flight," *Scientific American* (February 1998), pp. 38–47.

34. K. P. Dial, "Wing-Assisted Incline Running and the Evolution of Flight," *Science,* No. 299 (2003), pp. 402–4; P. Burgers and L. M. Chiappe, "The Wing of Archaeopteryx as a Primary Thrust Generator," *Nature,* No. 399 (1999), pp. 60–62; P. Burgers and Kevin Padian, "Why Thrust and Ground Effect Are More Important Than Lift in the Evolution of Sustained Flight," in J. Gauthier and L. F. Gall (eds.), *New Perspectives on the Origin and Evolution of Birds: Proceedings of the International Symposium in Honor of John H. Ostrum* (New Haven, Conn.: Peabody Museum of Natural History, 2001), pp. 351–61.

35. Alan Gishlick, "Evolutionary Paths to Irreducible Systems: The Avian Flight Apparatus," in Young and Edis (eds.), *Why Intelligent Design Fails,* pp. 58–71.

36. A. J. Spormann, "Gliding Motility in Bacteria: Insights from Studies of *Myxococcus Xanthus,*" *Microbiology and Molecular Biology Reviews* No. 63 (1999), pp. 621–41.

37. S. I. Aizawa, "Bacterial Flagella and Type-III Secretion Systems," *FEMS Microbiology Letters* No. 202 (2001), pp. 157–64.

38. Ian Musgrave, "Evolution of the Bacterial Flagellus," in Young and Edis (eds.), *Why Intelligent Design Fails,* pp. 72–84.

39. Dembski, *No Free Lunch,* pp. 159–60.

40. Ibid., pp. 212, 223.

41. Ibid., pp. 166–73.

42. Lynn Margulis and Dorion Sagan, *Acquiring Genomes: A Theory of the Origins of Species* (New York: Basic Books, 2002).

43. Richard Dawkins, "Weaving a Genetic Rainbow: How Evolution Increases Information in the Genome," *Skeptic* Vol. 7, No. 2 (2000), pp. 64–69.

44. This point was well made by Kenneth Miller in his book *Finding Darwin's God* (New York: Perennial, 2000).

45. Sean Carroll, "The Origins of Form," *Natural History* (November 2005), pp. 58–63; Sean Carroll, *Endless Forms Most Beautiful: The New Science of Evo Devo* (New York: W. W. Norton, 2005).

46. For examples see Douglas Futuyma, *Evolution* (Sunderland, Mass.: Sinauer, 2005).

47. Douglas Futuyma, "On Darwin's Shoulders," *Natural History* (November 2005), pp. 64–68.

48. Henry Morris, *The Troubled Waters of Evolution* (San Diego, Calif.: Creation Life, 1972), p. 110.

49. Peter Atkins, *The Second Law: Energy, Chaos and Form* (New York: W. H. Freeman, 1994).

50. Stuart Kauffman, *The Origins of Order: Self-Organization and Selection in Evolution* (Oxford: Oxford University Press, 1993).

51. Richard Hardison, *Upon the Shoulders of Giants* (Baltimore, Md.: University Press of America, 1985). Independently of Hardison, and around the same time, Richard Dawkins famously conducted the same computer experiment, as reported in his book *The Blind Watchmaker* (New York: W. W. Norton, 1986), except he used a different phrase—"Methinks it is like a weasel." Neither one of them knew about the other's program. Dawkins produced his program in 1984. There is no way he could have known about Hardison's work because it was not published in any form that would have been available to anyone but the students in our class. And Hardison didn't know about Dawkins's program. When Dawkins read about Hardison's program he queried me. I explained the origin of the coincidence, to which he responded:

> Thank you for clearing up the mystery. Yes, the coincidence is fascinating. But it is not all that surprising, and you have spotted that it makes a good lesson in paranormal debunking. Once one has grasped (from Darwin) the paramount importance of ratcheted *cumulative* selection when faced with the Argument from Statistical Improbability, one's thoughts naturally turn to the famous monkeys who have so often been used to dramatise that Argument. It becomes the obvious simulation to do, to get the point across to doubters. It can easily be done with a little BASIC program, and that is what both Hardison and I did, at what must have been almost exactly the same time, 1984 or 1985. As for the superficial details, those pesky monkeys have always typed Shakespeare. Hamlet is his most famous play. To Be or Not to Be is the most famous passage from that play. I would probably have chosen it myself, except that I thought the dialogue between Hamlet and Polonius on chance resemblances in clouds would make a neat intro: hence "Methinks it is like a Weasel."

When Hardison read Dawkins's reply in *Skeptic* Vol. 9, No. 4, he wrote me:

> Incidentally, I never felt that the TOBEORNOTTOBE example was entirely original with me. Bob Newhart, the comic, did a very nice skit in which he proposed an infinite number of monkeys working with an infinite number of typewriters, and then he realized that he would also need an infinite number of "inspectors" looking over the shoulders of the monkeys to see if anything meaningful occurred. Newhart then put himself into the role of one of these inspectors, spending another boring day and finding nothing. "Dum de dum de dum . . . Boring . . .

Oh . . . Hey, Charlie, I think I have one. Let's see, yeah. 'To Be Or Not
To Be, that is the acxrotphoeic.'" I simply realized that Bob's humor
might be a useful way of helping students to comprehend the selective
nature of the "struggle for survival." So you see that my contribution
was minimal.

52. Jonathan Wells, *Icons of Evolution: Science or Myth? Why Much of What
We Teach About Evolution Is Wrong* (Washington, D.C.: Regnery, 2000).

53. Stephen Jay Gould, "Abscheulich! (Atrocious!)," *Natural History* (March
2000).

54. Isaac Asimov, foreword to D. Goldsmith (ed.), *Scientists Confront Velikovsky*
(Ithaca, N.Y.: Cornell University Press, 1977), pp. 7–15. In his book *Worlds in Collision,* Immanuel Velikovsky proposed a radical theory of planetary history in which
the planets went careening through the solar system, impacting one another like so
many billiard balls, all in ancient human history and recorded in the myths of
peoples around the world, which became the primary data source for Velikovsky.

5. Science under Attack

1. Michael Shermer, "The Chaos of History," *Nonlinear Science Today* Vol. 2,
No. 4 (1993), pp. 1–13; "Exorcising LaPlace's Demon: Chaos and Antichaos, His-
tory and Metahistory," *History and Theory* Vol. 34, No. 1 (1995), pp. 59–83;
"Chaos Theory," in D. R. Woolf (ed.), *The Encyclopedia of Historiography* (New
York: Garland Publishing, 1996); "The Crooked Timber of History: History Is
Complex and Often Chaotic. Can We Use This to Better Understand the Past?"
Complexity Vol. 2, No. 6 (July–August 1997), pp. 23–29.

2. Michael Shermer, *Denying History* (Berkeley: University of California
Press, 2000).

3. Lynn Margulis, M. F. Dolan, and R. Guerrero, "The Chimeric Eukaryote:
Origin of the Nucleus from the Karyomastigonts in Amitochondriate Protists,"
Proceedings of the National Academy of Sciences No. 97 (2002), pp. 6954–59.
Lynn Margulis and Dorion Sagan, *Microcosmos: Four Billion Years of Microbial
Evolution* (Berkeley: University of California Press, 1997). Lynn Margulis, *Symbi-
otic Planet: A New Look at Evolution* (New York: Basic Books, 1998). Lynn Mar-
gulis and Dorion Sagan, *Acquiring Genomes: A Theory of the Origins of Species*
(New York: Basic Books, 2002).

4. Michael Shermer, *Why People Believe Weird Things* (New York: W. H. Freeman, 1997), pp. 18–19.

5. William R. Overton, "Memorandum Opinion of United States District
Judge William R. Overton in McLean v. Arkansas, 5 January 1982," in Langdon
Gilkey (ed.), *Creationism on Trial* (New York: Harper & Row, 1985), pp. 280–83.

6. The *amicus curiae* brief is both concise (at 27 pages) and well documented
(32 lengthy footnotes), and I discuss it at length in my book *Why People Believe
Weird Things,* pp. 154–72.

7. Stephen Jay Gould, "Knight Takes Bishop," *Natural History* (May 1986).

8. For a detailed account of the trial see: Burt Humburg and Ed Brayton, "Dover Decision—Design Denied: Report on *Kitzmiller et al. v. Dover Area School District*," *Skeptic* Vol. 12, No. 2 (2006), pp. 23–29. Court documents are related materials may be found at the Web page for the National Center for Science Education: http://www.ncseweb.org/.

6. The Real Agenda

1. William Dembski, *The Design Revolution: Answering the Toughest Questions about Intelligent Design* (Downers Grove, Ill.: InterVarsity Press, 2004), p. 41.

2. Quoted in Steve Benen, "Science Test," *Church & State* (July–August 2000). Available online at http://www.au.org/churchstate/cs7002.htm.

3. William Dembski, "Signs of Intelligence: A Primer on the Discernment of Intelligent Design," *Touchstone* (1999), p. 84.

4. Quoted in Benen, "Science Test," *Church & State* (July–August 2000).

5. Quoted in Jay Grelen, "Witnesses for the Prosecution," *World* (November 30, 1996). Available online at http://www.worldmag.com/world/issue/11-30-96/national_2.asp.

6. Wedge Document, Phase III. For an extensive discussion and reprinting of the Wedge Document see Barbara Forrest and Paul R. Gross, *Creationism's Trojan Horse: The Wedge of Intelligent Design* (New York: Oxford University Press, 2004).

7. Phillip Johnson, *The Wedge of Truth: Splitting the Foundations of Naturalism* (Downers Grove, Ill.: InterVarsity Press, 2000).

8. William Dembski, "Intelligent Design's Contribution to the Debate over Evolution: A Reply to Henry Morris," 2005. Available online at http://www.designinference.com/documents/2005.02.Reply_to_Henry_Morris.htm.

9. Dembski, *Design Revolution*, p. 319.

10. Paul Nelson statement available online at http://www.uncommondescent.com/index.php/archives/49#more-49.

11. Quoted in "By Design: A Whitworth Professor Takes a Controversial Stand to Show That Life Was No Accident. Stephen C. Meyer Profile," *Whitworth Today,* Whitworth College, Winter 1995. Available online at http://www.arn.org/docs/meyer/sm_bydesign.htm.

12. Jodi Wilgoren, "Politicized Scholars Put Evolution on the Defensive," *New York Times,* August 21, 2005. The Discovery Institute is not alone. In Virginia, Liberty University sponsored the Creation Mega Conference with a Kentucky group called Answers in Genesis, which raised $9 million in 2003 for their efforts to teach biblical Young Earth Creationism. See "Major Grants Increase Programs, Nearly Double Discovery Budget," Discovery Institute *Journal* (1999). Available online at http://www.discovery.org/w3/discovery.org/journal/1999/grants.html.

13. John Schwartz, "Smithsonian to Screen a Movie That Makes a Case against Evolution," *New York Times,* May 28, 2005.

14. Christoph Schönborn, "Finding Design in Nature," *New York Times,* July 7, 2005.

15. Bruce Chapman, "Ideas Whose Time Is Coming," Discovery Institute *Journal* (Summer 1996). Available online at http://www.discovery.org/w3/discovery.org/journal/president.html.

16. Wilgoren, "Politicized Scholars Put Evolution on the Defensive," *New York Times,* August 21, 2005.

7. Why Science Cannot Contradict Religion

1. Francis Darwin (ed.), *The Life and Letters of Charles Darwin,* 3 vols. (London: John Murray, 1887), Vol. 2, p. 105.

2. Francis Darwin (ed.), Ibid., Vol. 1, pp. 280–81.

3. Janet Browne, *Charles Darwin: A Biography* (New York: Knopf, 1995), p. 503. See also Adrian Desmond and James Moore's thoughtful discussion in their book, *Darwin* (New York: Warner Books, 1991), p. 387.

4. Letter to J. Fordyee reprinted in Gavin De Beer, "Further Unpublished Letters of Charles Darwin," *Annals of Science* 14 (1958), p. 88.

5. Charles Darwin letter to Edward Aveling, October 13, 1880, quoted in Desmond and Moore, *Darwin,* p. 645. See also Stephen Jay Gould, "A Darwinian Gentleman at Marx's Funeral," *Natural History* (September 1999).

6. The conflicting-worlds model of science and religion began in the late nineteenth century with the publication of two influential works that set the tone of the relationship for the next century: John William Draper's 1874 *History of the Conflict between Religion and Science* and Andrew Dickson White's 1896 *A History of the Warfare of Science with Theology in Christendom.* Both Draper and White presented simplified histories of the alleged war through such prominent events as the discovery of the earth's sphericity, Galileo's heresy trial, and the 1860 Huxley-Wilberforce debate over evolution, all of which historians of science have discovered had a considerably more nuanced history.

7. Pope John Paul II's definitive statements on the relationship of religion and science, faith and reason, are presented in two encyclicals: *Truth Cannot Contradict Truth* (1996) and *Fides et Ratio* (1998).

8. Stephen Jay Gould, "Nonoverlapping Magisteria," *Natural History* (March 1997). See also his expanded discussion in Stephen Jay Gould, *Rocks of Ages: Science and Religion in the Fullness of Life* (New York: Ballantine Books, 1999).

9. Karl Popper, *The Logic of Scientific Discovery* (New York: Basic Books, 1959), pp. 40–41.

10. R. Sloan, E. Bagiella, and T. Powell, *The Lancet* Vol. 353 (2000), pp. 664–67. Michael Shermer, "Flying Carpets and Scientific Prayer," *Scientific American* (November 2004), p. 35.

11. John Paul II, *Truth Cannot Contradict Truth.* Message to the Pontifical Academy of Sciences, 1996.

8. Why Christians and Conservatives Should Accept Evolution

1. Edward J. Larson and Larry Witham, "Scientists Are Still Keeping the Faith," *Nature* Vol. 386 (April 3, 1997), p. 435. The survey of 1,600 scientists was conducted by Elaine Howard Ecklund of Rice University. See Lea Plante, "Spirituality Soars among Scientists," *Science and Theology News* (October 2005), pp. 7–8.

2. If someone fully accepts the findings of science but privately believes that the forces of nature as described by science were God's way of creating the world and its inhabitants, I see no reason to go out of my way to object.

3. President Jimmy Carter's written statement, issued by the Carter Center on January 30, 2004, and reported widely in the media. See, for example, http://www.cnn.com/2004/EDUCATION/01/30/georgia.evolution/.

4. John Paul II, "Message to the Pontifical Academy of Sciences," reprinted in *The Quarterly Review of Biology* Vol. 72, No. 4 (December 1997), pp. 381–83.

5. Pew Research Center for People & the Press survey data available online at http://peoplepress.org/reports/display.php3?ReportID=254. Results for this survey were based on telephone interviews conducted under the direction of Princeton Survey Research Associates International among a nationwide sample of 2,000 adults eighteen years of age or older between July 7 and 17, 2005. Harris poll data available online at http://www.harrisinteractive.com/harris_poll/index.asp?PID=581.

The Harris poll was conducted by telephone within the United States among a nationwide cross section of 1,000 adults eighteen years of age or older between June 17 and 21, 2005.

6. Charles Darwin, *The Descent of Man, and Selection in Relation to Sex* (London: John Murray, 1871), Vol. 1, pp. 71–72.

7. T. H. Huxley, *Evolution and Ethics* (New York: D. Appleton and Co., 1894).

8. David M. Buss, *The Dangerous Passion: Why Jealousy Is as Necessary as Love and Sex* (New York: Free Press, 2002). See also David P. Barash and Judith E. Lipton, *The Myth of Monogamy: Fidelity and Infidelity in Animals and People* (New York: W. H. Freeman, 2001).

9. Paul Ekman, *Telling Lies: Clues to Deceit in the Marketplace, Marriage, and Politics* (New York: W. W. Norton, 1992); Paul Ekman, *Emotions Revealed: Recognizing Faces and Feelings to Improve Communication and Emotional Life* (New York: Times Books, 2003).

10. Adam Smith (R. H. Campbell and A. S. Skinner, gen. eds., W. B. Todd textual ed.), *An Inquiry into the Nature and Causes of the Wealth of Nations,* 2 vols. (Oxford: Clarendon Press, 1976), p. 14. Originally published in 1776.

11. Ibid., p. 423. Emphasis added.

12. Charles Darwin, *On the Origin of Species by Means of Natural Selection: or, The Preservation of Favoured Races in the Struggle for Life* (London: John Murray, 1859), p. 84. Emphasis added. The parallels between natural selection and the invisible hand are salient. Although Darwin does not reference Smith directly, when he matriculated at Edinburgh University for medical studies in October of 1825, he read the works of such great Enlightenment thinkers as David Hume, Edward Gibbon, and Adam Smith. A decade later, upon his return home from the five-year voyage around the world on the *Beagle,* Darwin revisited these works, reconsidering their theoretical implications in light of the new data he had collected. Darwin scholars are largely in agreement that he modeled his theory of natural selection after Smith's theory of the invisible hand, and there is a sizable literature on the connection between them. See, for example, Toni Vogel Carey, "The Invisible Hand of Natural Selection, and Vice Versa," *Biology & Philosophy*

Vol. 13, No. 3 (July 1998), pp. 427–42; Michael T. Ghiselin, *The Economy of Nature and the Evolution of Sex* (Berkeley: University of California Press, 1974); Stephen Jay Gould, "Darwin's Middle Road," in *The Panda's Thumb* (New York: W. W. Norton, 1980), pp. 59–68; Stephen Jay Gould, "Darwin and Paley Meet the Invisible Hand," in *Eight Little Piggies* (New York: W. W. Norton, 1993), pp. 138–52; Elias L. Khalil, "Evolutionary Biology and Evolutionary Economics," *Journal of Interdisciplinary Economics* Vol. 8, No. 4 (1997), pp. 221–44; Silvan S. Schweber, "Darwin and the Political Economists: Divergence of Character," *Journal of the History of Biology* Vol. 13 (1980), pp. 195–289.

9. The Real Unsolved Problems in Evolution

1. From Charles Darwin's diary. See R. D. Keynes (ed.), *Charles Darwin's Beagle Diary* (Cambridge, U.K.: Cambridge University Press, 1988), p. 353. I was able to snorkel in the bay and observe from beneath the waves the remarkable ability of the blue-footed boobies to penetrate several meters of water to nab their prey.

2. The conference was the brainchild of Carlos Montufar, the co-founder of the sponsoring institution—the Universidad San Francisco de Quito—and a reader of *Skeptic* magazine who invited me to speak on the evolution-creation controversy. The five-day conference (June 8–12) was hosted by the Galápagos Academic Institute for the Arts and Sciences (GAIAS), a high-tech facility flanked by low-tech homes and businesses. GAIAS is operated by the Universidad San Francisco de Quito, which obtained additional support from the U.S. National Science Foundation (which paid the way for graduate students in evolutionary biology to attend), Microsoft (which provided computers and Internet technology for GAIAS), UNESCO, and OCP Ecuador S.A., an oil conglomerate that provided additional funding.

3. Donald Rumsfeld quoted in Hart Seely, "The Poetry of D. H. Rumsfeld," Slate.com, April 2, 2003. Available online at http://slate.msn.com/id/2081042. See also the *New Yorker* article elaborating on Rumsfeld's souce for the quote: "Rumsfeld's work on the ballistic-missile commission convinced him that intelligence analysts were not asking themselves the full range of questions on any given subject—including what they didn't know. Rumsfeld gave me a copy of some aphorisms he had collected during the process of assessing the ballistic-missile threat. 'Some of these are humorous,' he said, not quite accurately. One was 'There are knowns, known unknowns, and unknown unknowns.' (The saying is attributed, naturally, to 'Unknown.') 'I think this construct is just powerful,' Rumsfeld said. 'The unknown unknowns, we do not even know we don't know them.'" Jeffrey Goldberg, "The Unknown: The C.I.A. and the Pentagon Take Another Look at Al Qaeda and Iraq," *The New Yorker*, February 10, 2003. Available online at http://www.newyorker.com/fact/content/articles/030210fa_fact.

4. Stephen Meyer, "The Origin of Biological Information and the Higher Taxonomic Categories," *Proceedings of the Biological Society of Washington* (June 2004). For an analysis of this paper and how it got published, see Robert Weitzel, "The Intelligent Design of a Peer-Reviewed Article," *Skeptic* Vol. 11, No. 4 (2005), pp. 44–48.

Epilogue: Why Science Matters

1. Richard Feynman, *Surely You're Joking, Mr. Feynman* (New York: W. W. Norton, 1985), p. 339.

2. Sagan quotes from the DVD edition of *Cosmos*; the opening quote is from DVD 1, scene 1, and the subsequent quotes are from DVD 13, scene 11. See also the book version of the documentary series: Carl Sagan, *Cosmos* (New York: Random House, 1980), pp. 4, 345.

3. Sagan's biographer, Keay Davidson, in fact, called Sagan's novel *Contact* "one of the most religious science-fiction tales ever written." Keay Davidson, *Carl Sagan: A Life* (New York: Wiley, 1999), p. 350. Consider what happens when the heroine of the story, Ellie Arroway (played by Jodie Foster in the film version), discovers that pi—the ratio of the circumference of a circle to its diameter—is numerically encoded in the cosmos, and that this is proof that a super-intelligence designed the universe: "The universe was made on purpose, the circle said. In whatever galaxy you happen to find yourself, you take the circumference of a circle, divide it by its diameter, measure closely enough, and uncover a miracle—another circle, drawn kilometers downstream of the decimal point. In the fabric of space and in the nature of matter, as in a great work of art, there is, written small, the artist's signature. Standing over humans, gods, and demons, subsuming Caretakers and Tunnel builders, there is an intelligence that antedates the universe." Carl Sagan, *Contact* (New York: Pocket Books, 1986), pp. 430–31.

4. Quoted in Richard Dawkins, *Unweaving the Rainbow* (New York: Houghton Mifflin, 1998), p. 40. Dawkins's own essay on the spiritual beauty of science is a classic in this genre. He writes, for example: "Science is poetic, ought to be poetic, has much to learn from poets and should press good poetic imagery and metaphor into its inspirational service." Dawkins then proceeds to do just that, in such elegant passages as this: "I believe that an orderly universe, one indifferent to human preoccupations, in which everything has an explanation even if we still have a long way to go before we find it, is a more beautiful, more wonderful place than a universe tricked out with capricious, *ad hoc* magic." Richard Feynman also expounded on the aesthetics of science: "The beauty that is there for you is also available for me, too. But I see a deeper beauty that isn't so readily available to others. I can see the complicated interactions of the flower. The color of the flower is red. Does the fact that the plant has color mean that it evolved to attract insects? This adds a further question. Can insects see color? Do they have an aesthetic sense? And so on. I don't see how studying a flower ever detracts from its beauty. It only adds." Richard Feynman, *What Do YOU Care What Other People Think?* (New York: Bantam Books, 1988.)

SELECTED BIBLIOGRAPHY

✦

A Reader's Guide to the Evolution–Intelligent Design Debate

Intelligent Design creationists are nothing if not prolific. Their arguments summarized in this book can be found in a number of works published over the past decade, the most prominent and widely quoted of which include:

Behe, Michael. *Darwin's Black Box: The Biochemical Challenge to Evolution.* New York: Free Press, 1996.

Campbell, John Angus, and Stephen C. Meyer, eds. *Darwinism, Design, and Public Education.* East Lansing: Michigan State University Press, 2003.

Davis, Percival William, and Dean Kenyon. *Of Pandas and People.* Dallas, Tex.: Haughton, 1993.

Dembski, William. *The Design Inference: Eliminating Chance through Small Probabilities.* New York: Cambridge University Press, 1998.

Dembski, William. *Intelligent Design: The Bridge between Science and Theology.* Downers Grove, Ill.: InterVarsity Press, 1999.

Dembski, William. *No Free Lunch: Why Specified Complexity Cannot Be Purchased without Intelligence.* New York: Rowman & Littlefield, 2002.

Dembski, William. *The Design Revolution: Answering the Toughest Questions about Intelligent Design.* Downers Grove, Ill.: InterVarsity Press, 2004.

Denton, Michael. *Evolution: A Theory in Crisis.* Bethesda, Md.: Adler and Adler, 1985.

Johnson, Phillip. *Darwin on Trial.* Downers Grove, Ill.: InterVarsity Press, 1991.

Johnson, Phillip. *Reason in the Balance: The Case against Naturalism in Science, Law, and Education.* Downers Grove, Ill.: InterVarsity Press, 1995.

Johnson, Phillip. *Defeating Darwinism by Opening Minds.* Downers Grove, Ill.: InterVarsity Press, 1997.

Johnson, Phillip. *The Wedge of Truth: Splitting the Foundations of Naturalism.* Downers Grove, Ill.: InterVarsity Press, 2000.

Wells, Jonathan. *Icons of Evolution: Science or Myth? Why Much of What We Teach about Evolution Is Wrong.* Washington, D.C.: Regnery, 2000.

Scientists and scholars began responding to Intelligent Design Creationism within a few years of the movement's rise to prominence. Here is a short list of books that most capably refute the arguments of Intelligent Design, as well as expose at greater length the political and religious agenda behind the movement:

Dawkins, Richard. *A Devil's Chaplain: Reflections on Hope, Lies, Science, and Love.* Boston: Houghton Mifflin, 2003.

Dawkins, Richard. *The Ancestor's Tale: A Pilgrimage to the Dawn of Evolution.* Boston: Houghton Mifflin, 2004.

Forrest, Barbara, and Paul Gross. *Creationism's Trojan Horse: The Wedge of Intelligent Design.* New York: Oxford University Press, 2004.

Miller, Kenneth. *Finding Darwin's God.* New York: Perennial, 2000.

Pennock, Robert. *Tower of Babel: The Evidence against the New Creationism.* Cambridge, Mass.: MIT Press, 1999.

Pennock, Robert, ed. *Intelligent Design Creationism and Its Critics.* Cambridge, Mass.: MIT Press, 2001.

Perakh, Mark. *Unintelligent Design.* Buffalo, N.Y.: Prometheus Books, 2004.

Pigliucci, Massimo. *Denying Evolution.* Cambridge, Mass.: Sinauer, 2002.

Ruse, Michael. *Darwin and Design: Does Evolution Have a Purpose?* Cambridge, Mass.: Harvard University Press, 2003.

Ruse, Michael. *The Evolution-Creation Struggle.* Cambridge, Mass.: Harvard University Press, 2005.

Scott, Eugenie. *Evolution vs. Creationism: An Introduction.* Berkeley: University of California Press, 2004.

Shanks, Niall. *God, the Devil, and Darwin: A Critique of Intelligent Design Theory.* New York: Oxford University Press, 2004.

Young, Matt, and Taner Edis, eds. *Why Intelligent Design Fails: A Scientific Critique of the New Creationism.* New Brunswick, N.J.: Rutgers University Press, 2004.

There are also substantial resources on the Internet regarding creationism and evolution:

Pro–Intelligent Design Web sites include:
Access Research Network: http://www.arn.org
Design Inference Web Site: http://www.designinference.com
Discovery Institute, Center for Science and Culture: http://www.discovery.org/csc
Evolution vs. Design: http://www.evidence.info/design/

God and Science: http://www.godandscience.org/evolution/
Intelligent Design and Evolution Awareness Club: http://www.ucsd.edu/~idea
Intelligent Design Network: http://www.intelligentdesignnetwork.org
Origins.org: http://www.origins.org/menus/design.html
Uncommon Descent: William Dembski's weblog: http://www
    .uncommondescent.com/

Pro-Evolution Web sites include:
Anti-Evolutionists: http://www.antievolution.org/people/
Biological Sciences Curriculum Study: http://www.bscs.org
Evolution Project: http://www.pbs.org/evolution
Evolution, Science and Society: http://evonet.sdsc.edu/evoscisociety
Institute for Biblical and Scientific Studies: http://bibleandscience.com
Institute on Religion in an Age of Science: http://www.iras.org
Metanexus Institute on Science and Religion: http://www.metanexus.org
National Center for Science Education: http://www.natcenscied.org
National Association of Biology Teachers: http://www.nabt.org
National Science Teachers Association: http://www.nsta.org
Skeptics Society: http://www.skeptic.com
Talk Design: http://www.talkdesign.org
Talk Origins forum: http://www.talkorigins.org
Talk Reason: http://www.talkreason.org

ACKNOWLEDGMENTS

◆

This book may have a single byline, but there is a team of people that have either supported this project in particular or my work in general through the Skeptics Society, *Skeptic* magazine, and *Scientific American*. At this venerable institution of American publishing, now over a century and a half old, I wish to thank my immediate editor, Mariette DiChristina, for her unparalleled ability to make my column readable each month, and John Rennie, for granting me the freedom to explore various regions of skepticism, as well as for his relentless defense of science in general and evolutionary theory in particular in the pages of *Scientific American*.

At the Skeptics Society and *Skeptic* magazine a giant debt of gratitude goes to Pat Linse for her continued efforts on behalf of science and skepticism, for her tireless good cheer during countless hours of working to get the magazine out and keep the ball moving down the field, and especially for her friendship and support. In various supporting roles in the society and magazine are office manager Tanja Sterrmann, office associate Sarah Lether, *Jr. Skeptic* editor and illustrator Daniel Loxton, web designer and director Emrys Miller and his associate William Bull at Rocketday Arts, *Jr. Skeptic* magician and science educator Bob Friedhoffer, videographer Brad Davies, photographer Dave Patton, senior editor Frank Miele, senior scientists David Naiditch, Bernard Leikind, Liam McDaid, and Thomas McDonough, artists Stephen Asma, Jason Bowes, Jean Paul Buquet, John Coulter, Janet Dreyer, and Adam Caldwell, editorial associates Gene Friedman, and Sara Meric, and Caltech lecture staff Diane Knudtson, Haime Botero, Michael Gilmore, Cliff Caplan, Tim Callahan, and Bonnie Callahan.

I would also like to recognize *Skeptic* magazine's board members: Richard Abanes, David Alexander, the late Steve Allen, Arthur Benjamin, Roger Bingham, Napoleon Chagnon, K. C. Cole, Jared Diamond, Clayton J. Drees, Mark Edward, George Fischbeck, Greg Forbes, the late Stephen Jay Gould, John Gribbin, Steve Harris, William Jarvis, Lawrence Krauss, Gerald Larue, William McComas, John Mosley, Richard Olson, Donald Prothero, James Randi, Vincent Sarich, Eugenie Scott, Nancy Segal, Elie Shneour, Jay Stuart Snelson, Frank Sulloway, Julia Sweeney, Carol Tavris, and Stuart Vyse.

As always, I wish to thank my agents, Katinka Matson and John Brockman, for the always professional manner in which they treat the literary business, as well as to acknowledge Paul Golob at Henry Holt / Times books, who oversaw the project, and most notably Robin Dennis, my editor, whose opinions on literary matters I trust more than my own. Jessica Firger in the Holt publicity department has unfailingly supported our long-range mission of promoting science and critical thinking by reaching larger audiences, and for this I am deeply grateful. I also thank Emily DeHuff for sharp-eyed copy editing of the manuscript, Lisa Fyfe for the creative cover design, Victoria Hartman for the elegant interior design, and Rita Quintas for the editorial production process.

Thanks as well go to David Baltimore, Kip Thorne, Christof Koch, Susan Davis, Chris Harcourt, and Ramanuj Basu at Caltech for their continued support of the Skeptics Science Lecture Series at the institute. Larry Mantle, Ilsa Setziol, Jackie Oclaray, Julia Posie, and Linda Othenin-Girard at KPCC 89.3 FM radio in Pasadena have been good friends and valuable supporters for promoting science and critical thinking on the air. Robert Zeps, John Moores, Thomas Glover, Robert Engman, Gerry Ohrstrom, and Glenn Camni have been especially supportive of the Skeptics Society, and to them I am especially appreciative.

Finally, I acknowledge Kim and Devin for being my family and all that that means—which is everything; and most notably for this book, special thanks go to Frank J. Sulloway, who has taught me more about science and evolution than I could learn from a library of books, and whose influence is reflected in the dedication of this book.

INDEX

✦

ABOUT THE AUTHOR

✦

MICHAEL SHERMER, PH.D., is the founding publisher of *Skeptic* magazine (www.skeptic.com), the executive director of the Skeptics Society, a monthly columnist for *Scientific American,* the host of the Skeptics Distinguished Science Lecture Series at the California Institute of Technology (Caltech), and the co-host and producer of the thirteen-hour Family Channel television series *Exploring the Unknown.* About Dr. Shermer, the late Stephen Jay Gould wrote: "As head of one of America's leading skeptic organizations, and as a powerful activist and essayist in the service of this operational form of reason, [he] is an important figure in American public life."

Shermer is the author of numerous books. He has written a trilogy on belief: the bestselling *Why People Believe Weird Things,* on pseudoscience, superstitions, and other confusions of our time; *How We Believe: Science, Skepticism, and the Search for God,* on the origins of religion and belief in God; and *The Science of Good and Evil: Why People Cheat, Gossip, Share, Care, and Follow the*

Golden Rule, on the evolutionary origins of morality. He has also published two collections of essays, *Science Friction: Where the Known Meets the Unknown,* about how the mind works and how thinking goes wrong, and *The Borderlands of Science,* which maps the fuzzy land between science and pseudoscience. He is also the author of a biography, *In Darwin's Shadow,* about the life and science of the co-discoverer of natural selection, Alfred Russel Wallace, and *Denying History,* on Holocaust denial and other forms of pseudohistory.

Shermer earned his B.A. in psychology from Pepperdine University, his M.A. in experimental psychology from California State University at Fullerton, and his Ph.D. in the history of science from Claremont Graduate University. He was a college professor for twenty years, teaching psychology, evolution, and the history of science at Occidental College; California State University, Los Angeles; and Glendale College. He lives in Southern California.